Funding for this

project was provided

by

John G. Walker

and

Roxann Moore

of

Coaldale, Colorado

Vol. IX.] 1891. [No. 1.

JOURNAL

OF THE

UNITED STATES ASSOCIATION

OF

CHARCOAL IRON WORKERS

Devoted to Iron Metallurgy and the Mining of Iron Ores.

Published for the Proprietors
by
JOHN BIRKINBINE, Editor,
No. 25 North Juniper St., Philadelphia, Pa.

Subscription price, three dollars per volume of six numbers (including index to volume). Single copies, fifty cents each.

To meet the demand for complete volumes of the Journal, the publishers have depleted the stock of certain numbers, so that it is considered advisable to direct attention to the fact that unless additional copies of some of the issues can be obtained, but few more complete sets can be supplied at the prices named in the advertisements. The numbers which at present are scarce and for which compensation in exchanges or in money will be paid are:

Vol. I., No. 1, June, 1880.
Vol. I., No. 2, August, 1880.

The Editor.

A Visit to the Hanging Rock Region of Ohio.

During a late trip through Southern Ohio we enjoyed a visit to the Olive furnace, located in the northern portion of Lawrence county, Ohio, which we believe is the only active blast-furnace in the United States, the stack of which is largely hewn from the solid rock.

We regret that it was impossible to renew acquaintance with the many friends in the Hanging Rock Region, who entertained the Association during its visit in connection with the annual meeting in 1881, a narrative of which appears in the JOURNAL, vol. ii., pp. 262–276; but during our short sojourn we were impressed with the fact that at present there are not more than ten charcoal blast-furnaces in the Hanging Rock district of Ohio, which may fairly be considered as being on the list of active plants, an equal number having within a decade suspended operations from various causes. Most of those which may be ranked as active, however, give promise of continuing as producers, having ample woodlands and ore-supplies.

The development of the West Virginia and Kentucky coking coal fields, and the construction of railroad connections to them is looked upon by those interested as promising much for the Hanging Rock Region in the near future. With the local ores as a basis and the opportunity of enriching the furnace-burden by lake ores, with Southern coke to compete with that from Pennsylvania, and the possibility of utilizing in part some of the local mineral fuel, it is prophesied that the Hanging Rock Region will be in a position to take advantage of its natural geographical location close to important markets which are accessible by both rail and water transportation.

We shall refer to but two of the charcoal furnaces at present, confining our notes to the description of the hewn stack of the Olive furnace, and the radical changes in the construction of the Hecla furnace; we shall, however discuss some of the features of the Hanging Rock district and its ore-supply.

The following table exhibits the amount of pig-iron made with charcoal, the amount made with bituminous coal or with a mixtur of bituminous coal and coke, and the total output of the blast-furnaces of the Hanging Rock Region of Ohio for a series of years.

PIG-IRON PRODUCTION OF THE HANGING ROCK REGION OF OHIO.

Years.	Net tons. Made with Charcoal.	Net tons. Made with Bituminous Fuel.	Net tons. Total.
1872	87,440	23,169	110,609
1873	92,365	28,601	120,966
1874	85,873	26,015	111,888
1875	57,413	36,899	94,312
1876	42,822	44,260	87,082
1877	40,212	44,544	84,756
1878	33,513	31,137	64,650
1879	43,445	43,097	86,542
1880	64,854	60,316	125,170
1881	61,487	77,500	138,987
1882	55,546	77,364	132,910
1883	38,134	82,455	120,589
1884	24,880	64,781	89,661
1885	18,018	68,837	86,855
1886	16,161	116,398	132,559
1887	17,244	126,487	143,731
1888	21,864	106,852	128,716
1889	22,467	84,737	107,204

The above demonstrates a moderate advance, except lately in the output of pig-iron made with mineral fuel, and a marked decrease in the output of the charcoal blast-furnaces.

THE OLIVE FURNACE

was constructed and first operated in the year 1846, and (with the exception of one year, when the death of a partner demanded a temporary cessation of operations) it has been in blast every year since its completion.

Originally the furnace was run with cold-blast, the pig-iron being consumed principally by charcoal forges along the Ohio River, for the production of blooms to be used in the manufacture of boiler plates. After being operated in this way for ten years, during which time the output averaged about six to eight tons per day, a hot-blast of the ring pattern was added, and subsequent improvements have changed this to a Player stove of 18 U-pipes; the original steam machinery was, about 1860, replaced by a horizontal engine with steam cylinder, 16 inches in diameter and 6 feet stroke, driving through gearing two horizontal blowing cylinders, 43 inches diameter and 5 feet stroke. This machinery is still in use.

Throughout the various changes the original stack built in 1846 was retained, and beyond an addition of 4 feet to the top masonry, making the total height 40 feet and widening one tuyere arch, the main structure is as it was first built.

The peculiarity of construction consists of the lower half of the blast-furnace stack being hewn from a ledge of solid rock, which is exposed for a height of over 20 feet. This rock ledge is a sandstone which cuts easily when first hewn, but hardens upon exposure. A suitable site having been selected, the earth in front of the rock was removed, and the entire rock face thus exposed was dressed for a length of 60 feet, and from this space the blast-furnace stack was cut, the additional height of stack, 20 feet, being obtained by masonry built upon the rock.

In general appearance and outline the lower portion of the stack of the Olive furnace does not, except in a total absence of bracing, differ materially from the older stacks which were built of stone masonry; the stack is square in plan with sloping sides, the bottom being 48 feet square and the top (20 feet above the bottom) 36 feet square. It stands at the sides and rear from 4 to 6 feet and at the top still further away from the mass of rock of which it originally formed a part. It is cored out to accommodate the lining of the furnace, and the sides are pierced for one fore arch and two tuyere arches, all cut in the solid rock. To illustrate the method of construction, the accompanying sketches were made. Fig. 1 is a front elevation of the Olive furnace stack; *A, B,—C, D* shows the original face of the ledge of rock, and *E, F,—G, H,* the portion of the ledge left standing for the furnace stack; the side pass-

FIG. 1. Elevation.

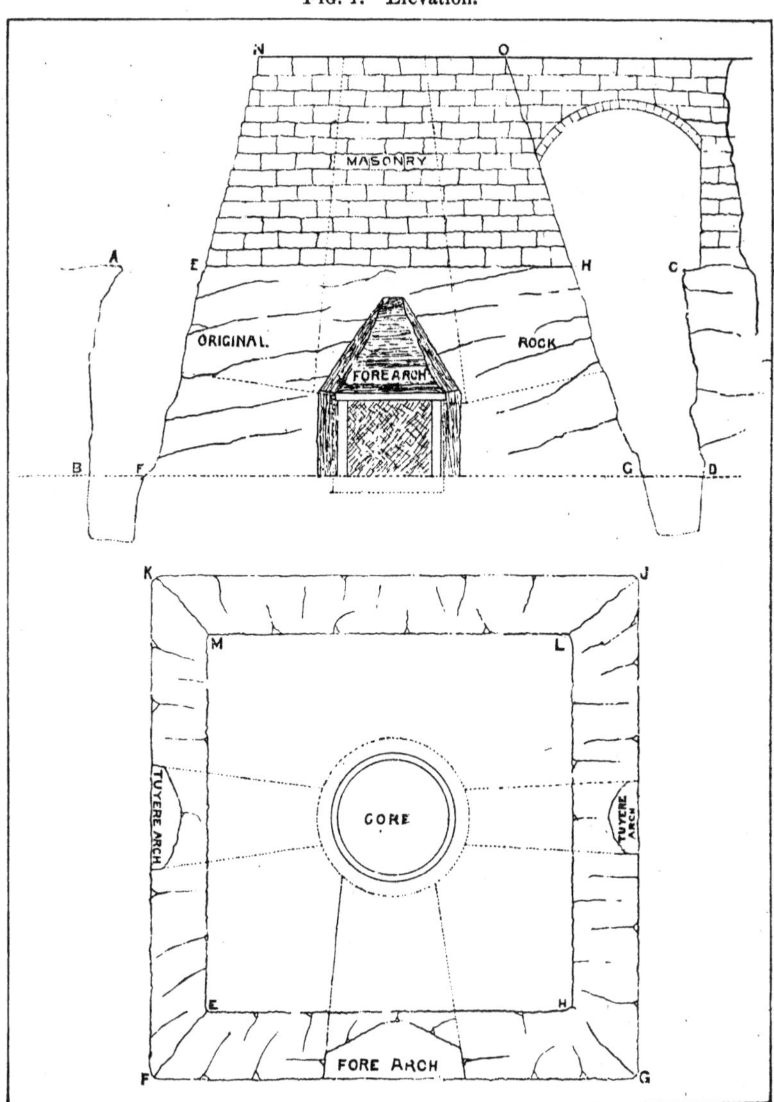

FIG. 2. Plan.

ages, *A*, *B*,—*F*, *E*, and *C*, *D*,—*G*, *H*, are cut back from the face over 50 feet, and a similar passage is formed at the rear of the stack. On the rock base of the stack the masonry, *N*, *E*, *H*, *O*, is placed. Fig. 2 shows a plan of the original rock base of the stack, with the upper or masonry portion removed; *E*, *H*, *M*, *L*, shows the plan of the top of the rock base, and *F*, *G*, *J*, *K* the dimensions of the rock base at hearth level; the three circles show the dimensions of the core cut in the rock for the furnace lining; the middle circle indicates the opening at the top of the rock base, the smaller circle shows the core in which the hearth bottom and crucible lining are placed and the larger dotted circle represents the outer diameter of what in a built-up stack would be the ring wall on which the shaft lining rests; the horizontal position and general dimensions of fore arch and tuyere arches are shown by dotted lines. In the elevation, Fig. 1, the core in the rock and the vertical position of the tuyere arch are shown by dotted lines, the fore arch being exhibited in elevation; the arch spanning the opening, *H*, *C*, carries the boilers, which, as is usual in the Hanging Rock Region, are placed over the tunnel head; beyond the boilers is the hot-blast stove. This arrangement of boilers, etc., which prevails in the Hanging Rock Region, is shown by plate opposite page 314, vol. 2, of JOURNAL.

Originally the inwall was laid with sandstone, and the hearth and bosh was constructed of the same material. In later years, however, fire-brick has been used for the inwall and lining; but hearth and bottoms are still cut from the sandstone, which occurs in great quantity, and of excellent quality, convenient to the Olive furnace.

The frequent changes of temperature have caused the mass of rock forming the lower portion of the stack to develop a number of cracks. At first these were small, but at each blast they have opened until they exhibit in a remarkable manner the marvellous power of expansion. This portion of the stack being without bands or buck-staves, has an appearance of weakness, which, however, on closer examination, is removed, and it is probable that the old hewn stack is good for years of service. As, however, it is evidently the last of its kind, we have placed on record these sketches of it, as a matter of interest to our readers.

Since the hot-blast was added to the furnace, the pig-iron has been employed chiefly for the manufacture of car-wheels, the charge being composed of what is locally known as red limestone ore, charcoal and limestone. The red limestone ore is a carbonate which has been in most places weathered to a limonite, and which is obtained by benching around or by driving into the neighboring hills. This ore lies immediately upon a limestone strata, which in turn rests upon a seam of bituminous coal of good quality, 4 feet thick, which latter is worked to a limited extent only.

On the Olive furnace tract, and on the adjoining Buckhorn furnace tract, the ore is obtained by benching, by stripping, and by underground exploitations. The benching was the original method of winning the ore, and at the proper elevation on all of the hills throughout these properties the position of the ore seam is noticed by a terrace or bench formed of the earth stripped from the ore, until handling of the earth was, by reason of the increased thickness, considered too expensive. It would appear, however, that this method of obtaining the ore could be economically carried further by the use of steam excavators to handle the overlying earth. Some of the ore is now won by stripping the earth from considerable areas in the bottoms where the ore approaches close to the surface, but most of the present operation is by underground mining, to facilitate which, the Messrs. McGugin have in use an air compressor and rock drills.

Close to the Olive furnace, and at other points in the Hanging Rock Region of Ohio, the occurrence of coal, limestone, and ore is such that the three materials could be taken from one opening, the overlying sand-rock forming a good roof for the mine. There seems good reason to anticipate that this course will be pursued when thorough tests determine the fuel value of the coal and improved transportation facilities make it accessible to consumers. We shall refer to the iron-ores of this region at the close of the article.

The charcoal is all made in meilers, and it and the ore are hauled in wagons drawn by ox-teams, five yoke to a team. These teams haul 4 tons of ore, or 185 bushels of charcoal to the furnace. The charcoal is rated by measure, the Association standard (2748 cubic inches) being used. Mr. W. N. McGugin, President of the United States Association of Charcoal Iron Workers, and his son, W. H.

McGugin, who operate Olive furnace, were instrumental in having this standard adopted by the Ohio Legislature.

The wood is largely hard, oaks and maples predominating, and from the 18,000 acres connected with the property, first and second growth timber is cut, making a very satisfactory character of charcoal. The furnace produces 15 to 16 tons of pig-iron per day, running but six days per week, the practice of the furnaces in the Hanging Rock Region of suspending operations from midnight Saturday to midnight Sunday being followed.

All of the ore used at the furnace is roasted with charcoal braize in kilns, circular in section and slightly conical. These kilns, two in number, have a total height of 21 feet 6 inches, the lower 2 feet 6 inches being taken up by a series of columns supporting a mantle on which rests a wrought-iron shell, 14 feet in diameter at the bottom, 13 feet in diameter at the top, and 19 feet high. This shell is lined with 9 inches of brick, between which and the plate-iron is sand backing. Similar kilns are used at other Hanging Rock furnaces, but the usual practice is to roast the ore in open heaps with charcoal braize or with bituminous coal slack.

The following analyses are of ores obtained from the property of the Olive Furnace, the determinations being made by the Geological Survey of Ohio, and were given us by Messrs. McGugin.

	A.	B.	C.	D.	E.	F.	G.
Iron,	40.01	38.02	37.603	49.23	48.06	55.085	60.89
Lime,	3.09	4.95	———	tr.	3.40	0.4	0.42
Silicia,	1.94	3.54	1.08	4.89	2.40	2.643	1.437
Manganese,	1.63	1.41	0.857	1.70	1.06	1.029	1.068
Sulphur,	0.65	0.03	0.554	0.15	0.029	0.02	0.028
Phosphorus,	0.55	0.28	0.132	0.51	1.02	0.325	0.356

Analyses A, B and C represent the raw blue limestone-ore, which carries a considerable amount of carbonic acid, the ore being practically a carbonate. Analyses D, E and F are determinations of the red limestone ore, which is really an altered carbonate, presenting many characteristics of limonite; and G is an analysis of the roasted-ore, the raw-ore being the same as shown in analysis F. Samples F and G show iron above what is usually found, and may be considered as picked specimens. The result of an entire blast at the Olive Furnace, in which the red-ore was largely used, showed that the furnace yield, after the ore had been roasted, was slightly above 50 per cent., *i.e.*, not quite two tons of roasted ore was required to produce a ton of pig-iron.

Concerning the limestone-ore, so-called, we quote from Vol. V., *Geological Survey of Ohio*: "This is the seam from which more than 50 furnaces of Ohio and the adjacent district of Kentucky have obtained, and many of which are still obtaining, their chief ore supply, some of them through two or even three scores of years. The iron made from it is the standard iron of the Ohio Valley for many uses; for strength and chilling qualities it is at the head of the list, a considerable amount of it being used in car-wheels and in machine castings. The Ferriferous limestone that bears it, is itself of great value and importance to the whole district, yielding almost the entire supply of furnace flux and lime.

"The ores that are known as *limestone-ores* present two distinct phases. The name is mainly confined to the Hanging Rock district of Southern Ohio, where it is applied to one well-known and very valuable seam, viz., the ore borne by the Ferriferous limestone of the general scale. The designation "limestone-ore" is specific in this portion of the State, being exclusively applied to this seam. The ore overlies the limestone, sometimes graduating insensibly into it, and sometimes separate and distinct from it, but lying in close proximity. Above the main sheet of ore, kidneys are generally to be found scattered through the clay. A common and characteristic form of the unaltered ore consists of grains of carbonate of iron, buried in a siliceous clay, this form does not blend with the limestone. The most valued form of the ore, especially for use in charcoal furnaces, is the limonite that has resulted from the weathering and transformation of the original carbonate. These ores, while originally carbonates of iron in every case, have been transformed along their lines of outcrop, and often under considerable cover, into hydrated peroxides or limonites. In many instances the transformation has been very thorough, the form, volume, specific gravity, texture and color of the ore being changed in the process. The change is always in the line of improvement in the quality of the ore."

This description of the ore holds good, not only for that on the Olive Furnace tract, but also on much of the territory recognized as the Hanging Rock region.

The Olive furnace is an example of which there are fewer now than in former times, of a property producing within its limits practically all the materials necessary for the construction and maintenance of the industry located upon it; an abundance of sandstone, brick, clay, fire-clay, limestone and timber supply building materials; the ore is mined, the flux quarried and the charcoal made on the property, while such incidentals, as pig-bed sand, coal for household use, etc., are obtained close to the furnace.

Mr. W. N. McGugin has been associated with the Olive furnace for nearly three decades, and has in later years combined the properties of the Olive and Buckhorn furnaces, the latter, a built-up stone stack, being at present inactive.

HECLA FURNACE.

The Hecla Iron and Mining Company has dismantled the old plant which was erected in 1834, and has just completed a new plant The stack is lined to 10 feet 4 inches bosh diameter, and has a height of 53 feet. It has a crucible 8 feet high, 53 inches diameter at bottom, 60 inches at top. It is blown with three tuyeres and is equipped with bell and hopper; the only charcoal blast-furnace in the Hanging Rock Region with closed top. Blast is supplied by two vertical blowing-engines, with blowing cylinders 60 inches diameter and 4 feet stroke, and heated in one 18 U-pipe Player oven. It will continue to use the local red and gray limestone ores and charcoal. The new furnace stack is an iron shell, supported on columns, with boilers and hot oven on ground. The shell is of sufficient diameter to accommodate a bosh of 15 feet, and is lined throughout with fire-brick.

Iron and Steel Manufacture in 1889.

THE annual report of the American Iron and Steel Association for the calendar year 1889 opens by an impressive table which shows the production for the past four years of the leading articles of iron and steel, and illustrates the remarkable growth of some of the special industries.

In the table are included the Spiegeleisen and ferro-manganese which were made in this country; the amount of 47,982 net tons made in 1886 was increased in 1888 to 54,769 tons, but in 1889 this special product reached a total of 85,823 net tons.

	Net tons.
Taking first pig-iron we find the output of the year 1886 was,	6,365,328
Which was increased in the year 1887 by,	821,878
And there was added in the year 1888,	81,301
In the year 1889 this amount was augmented by,	1,247,572
Making the output for the year 1889,	8,516,079
And showing a development of the pig-iron industry in three years of,	2,150,751
or 33.8 per cent increase on the output of 1886.	

In connection with these figures we present those lately issued by the Hon. R. P. Porter, Superintendent of the Eleventh Census, which vary somewhat from those of the American Iron and Steel Association, as the latter were collected for the calendar year 1889 and the former for the year ending June 30, 1880. Dr. William M. Sweet, to whom this branch of the census work was committed, says:

"The production of pig-iron during the year ended June 30, 1890, was the largest in the history of the iron industry of this country, amounting to 9,579,779 tons of 2000 pounds, as compared with 3,781,021 tons produced during the census year 1880 and 2,052,821 tons during the census year 1870. From 1870 to 1880 the increase in production amounted to 1,728,200 tons, or nearly 85 per cent., while from 1880 to 1890 the increase was 5,798,758 tons, or over 153 per cent. The following table shows the production of pig-iron in the various sections of the country in the census years 1870, 1880 and 1890, in tons of 2000 pounds, including castings made direct from the furnace. The statistics for 1870 and 1880 are for the census years ended May 31st, but for 1890 they cover the year ended June 30th.

From the table it will be seen that the pig-iron industry of New England has been practically stationary during the past twenty years, while during the same period, and especially since 1880, there has been a wonderful development of the manufacture of pig-iron in all other sections of the country.

In the census year 1880 the pig-iron industry was located in 24

States and 1 territory, but pig-iron was produced in only 22 States, the furnaces in Minnesota, North Carolina, and Utah being idle in that year. In the census year 1890 there were 25 States which contained completed blast-furnaces, and pig-iron was made in that year in each of these States, except Minnesota. Since 1880 the manufacture of pig-iron has been abandoned in Vermont and Utah, and during the same period 2 States, viz., Colorado and Washington, have engaged in its production. In the census year 1880 Minnesota contained one charcoal furnace, but it was not in operation in that year, and was abandoned in 1888. In the census year 1890 a large coke furnace was under construction in that State, but it was not completed until near the close of the year, and consequently had no product to report. California does not appear among the pig-iron producing States in either 1880 or 1890. A charcoal furnace was completed and put in operation in that State in 1881, but it has made no pig-iron since 1886, and is practically abandoned.

PIG-IRON PRODUCTION BY DISTRICTS.

Districts.	TONS OF 2000 POUNDS.		
	Year ended May 31, 1870.	Year ended May 31, 1880.	Year ended June 30, 1890.
New England States	34,471	30,957	33,781
Middle States	1,311,649	2,401,093	5,216,591
Southern States	184,540	350,436	1,780,909
Western States	522,161	995,335	2,522,351
Far Western States	....	3,200	26,147
Total	2,052,821	3,781,021	9,579,779

The relative rank of the various States is seen to have undergone many changes since 1880. Pennsylvania still retains its leadership as the producer of about one-half of the pig-iron that is annually made in the United States, producing 51 per cent. of the total production in the census year 1880 and over 49 per cent. in 1890. Ohio was second in rank in both 1880 and 1890, the output of pig-iron in the former year being over 14 per cent. of the total production in the United States, and in the latter year over 13 per cent. Alabama, which occupied tenth place in 1880, with an output of 62,336 tons, is now the third largest producer of pig-iron, the production of this state in 1890 amounting to 890,432 tons, an increase of more than 1328 per cent. over the production of 1880. Illinois, which was seventh in rank in 1880, is fourth in 1890; and New York, which was third in 1880, occupies fifth place in 1890. Virginia, which was seventeenth in rank in 1880, is now sixth; while Tennessee has gone from thirteenth to seventh place.

The following table shows the production of pig-iron in the United States, in tons of 2000 pounds, in the census years 1880 and 1890, arranged according to the fuel used, with the percentage of increase or decrease in production in 1890:

Fuel used.	Year ended May 31, 1880.	Year ended June 30, 1890.	Percentage of increase in 1890.	Percentage of decrease in 1890.
	Tons.	Tons.		
Anthracite alone	1,112,735	323,258	....	70.95
Mixed anthracite coal and coke	713,932	1,879,098	163.20	....
Coke and bituminous coal	1,515,107	6,711,974	343.00	....
Charcoal	435,018	655,520	50.69	....
Castings direct from furnace	4,229	9,929	134.78	....
Total	3,781,021	9,579,779	153.36	....

The production of Bessemer pig-iron in the United States during the census year 1890, which is included in the figures of total production of pig-iron, amounted to 4,233,372 tons. Of this quantity, Pennsylvania made 2,567,813 tons; Illinois, 616,659 tons; Ohio, 526,654 tons; New York, 174,574 tons; West Virginia, 101,178 tons; Maryland, 77,754 tons; Missouri, 68,629 tons; Wisconsin, 43,728 tons; New Jersey, 41,479 tons, and all other states a total of 14,904 tons.

The utilization of Bessemer steel for other purposes than rails is evident from the following statement of the production of Bessemer steel ingots and Bessemer steel rails:

In 1886.—1,763,667 tons of rails were produced from 2,541,493 tons of ingots, representing 69.4 per cent. of the weight of ingots made.

In 1887.—2,354,132 tons of rails were produced from 3,288,357 tons of ingots, representing 71.6 per cent. of the weight of ingots made.

In 1888.—1,552,631 tons of rails were produced from 2,812,500 tons of ingots, representing 55.2 per cent. of the weight of ingots made.

In 1889.—1,691,264 tons of rails were produced from 3,281,829 tons of ingots, representing 51.5 per cent. of the weight of ingots made.

The same holds true of open-hearth steel, for we find that the production of ingots has increased from 245,250 tons in 1886 to 419,488 tons in 1889, of which amounts but 5255 tons in 1886, and 3346 tons in 1889, were made into rails. Little variation is noted in the output of crucible steel, the product being 80,609 tons in 1886, and 84,969 tons in 1889.

In the manufacture of iron rails there has been a continual decline from 23,679 tons in 1886 to 10,258 tons in 1889.

The utilization of the excess of Bessemer and open-hearth steel ingots is also emphasized by the following figures of the production of

	1886. Net tons.	1887. Net tons.	1888. Net tons.	1889. Net tons.
Rolled iron, except rails	2,259,943	2,565,438	2,397,402	2,576,127
Rolled steel, " "	800,000	902,156	1,201,885	1,584,364

These indicate that new uses for steel are being found without (up to the present) displacing iron, for we note that although the amount of rolled steel in 1889 was 98 per cent. greater than in 1886, the quantity of rolled iron in each of the three years, 1887, 1888, 1889, was greater than in 1886.

Another indication of the employment of steel is exhibited by the following table, showing the production of iron and steel nails in the four years.

These figures show that the advances made by steel nails is by the displacement of iron nails, but they also show that the total production of nails was greater in 1886 and 1887 than in 1889.

	1886. Net tons.	1887. Net tons.	1888. Net tons.	1889. Net tons.
Kegs of iron cut nails..	5,191,984	3,419,578	2,170,107	1,778,082
Kegs of steel cut nails	2,968,989	3,489,292	4,323,484	4,032,676
Kegs of steel wire nails	600,000	1,250,000	1,500,000	2,200,000
Total..................	8,760,973	8,158,870	7,993,591	8,010,758

Before we leave this subject we would direct attention to the continued decline in the manufacture of blooms made with charcoal. The amounts in the four years were as follows:

	1886. Net tons.	1887. Net tons.	1888. Net tons.	1889. Net tons.
Pig, scrap, and ore blooms, made with charcoal..........	41,909	43,306	39,875	36,260

The production of iron blooms and billets direct from the ore in 1889 was 12,407 net tons, against 14,088 tons in 1888 and 15,088 tons in 1887. The production of blooms and billets from pig and scrap iron in 1889 was 23,853 net tons, against 25,787 tons in 1888 and 28,218 tons in 1887. The production of wrought iron direct from the ore in forges is now practically confined to the Lake Champlain district of New York, which produced 12,397 net tons in 1889, the few remaining tons being made in Tennessee. Of the pig and scrap blooms made last year, Pennsylvania produced 18,504 net tons, the remainder being made in New Jersey and Maryland.

In discussing the consumption of pig-iron and Bessemer steel rails in the United States, Mr. Swank gives the following figures:

APPARENT CONSUMPTION.

Years.	Pig Iron. Gross tons.	Bessemer Steel Rails. Gross tons.
1887	6,808,386	2,239,492
1888	6,688,744	1,449,294
1889	7,798,337	1,516,259

Our consumption of pig-iron since 1874, with an allowance in some years for foreign stocks and domestic exports, has been as follows:

APPARENT YEARLY CONSUMPTION OF PIG-IRON.

Years.	Gross tons.	Years.	Gross tons.
1874	2,500,000	1882	4,963,278
1875	2.000,000	1883	4,834,740
1876	1,900,000	1884	4,229,280
1877	2,150,000	1885	4,348,844
1878	2,500,000	1886	6,191,354
1879	3,432,534	1887	6,808,386
1880	3,990,415	1888	6,688,744
1881	4,982,565	1889	7,798,337

Our consumption of pig-iron thirty-five years ago was only one-tenth of what it now is.

Our pig-iron imports and exports for two years were as follows:

Gross tons of 2240 pounds, imported during 1888, 197,237 tons. Value, $3,007,327.

Gross tons of 2240 pounds, imported during 1889, 142,230 tons. Value, $2,863,137.

We exported 13,573 gross tons of pig-iron in 1889, against 14,364 tons in 1888, the latter being the largest quantity that we have exported for many years.

Concerning the capacity of the blast-furnaces of the United States, Mr. Swank says: "The annual capacity of our completed furnaces in November, 1887, was 10,990,993 net tons, and the capacity of the completed furnaces in November, 1889, was 13,168,233 net tons; an increase of 2,177,240 net tons, or 1,943,964 gross tons, in two years. Our production of pig-iron in the two years from 1887 to 1889, notwithstanding the extraordinary product of 1889, increased only 1,328,873 net tons. There was, therefore, a much greater increase in our blast-furnace capacity from 1887 to 1889 than in the production of pig-iron. On the 1st of November last twenty-seven new furnaces were then in the course of erection, and in January of the present year work had been commenced on ten more furnaces in the West and South. We shrink from any enumeration of the new blast-furnace enterprises which have been undertaken or projected since January. The fact must be sufficiently evident to our readers, however, that we now have a blast-furnace capacity which, after making due allowance for the fact that it cannot all be utilized at the same time, is far in advance of present requirements."

At the close of 1889 the total number of furnaces in the United States which were active, or likely to be some day active, was 570 and 32 new furnaces were in course of erection; 14 furnaces were completed during 1889, and 33 were burned, abandoned, or torn down to make room for new stacks.

The following table shows the number of furnaces in blast at the close of each year since 1885, classified according to the fuel used:

Kind of Fuel used.	1885.	1886.	1887.	1888.	1889.
Bituminous coal and coke	111	143	147	156	177
Anthracite and anthracite and coke	105	125	118	105	104
Charcoal	60	63	74	71	63
Total	276	331	339	332	344

Quoting again from the Census Bulletin we learn that notwithstanding the fact that the production of pig-iron has increased from 3,781,021 tons of 2000 pounds in 1880 to 9,579,779 tons in 1890, the total number of completed furnaces has decreased during the ten years from 681 to 562. Many furnaces which were in active list in 1880 have since been abandoned owing to their inabilty to profitably compete with the larger, better located, and more modern furnaces of the present day. The majority of these abandoned furnaces were of small capacity, and were able to produce and market pig-iron only during periods of great demand and consequent high prices, while the large number of new and improved furnaces which have been built during recent years, and which are favorably located for the supply of materials at low cost and within easy access to market, have now made the operation of these antiquated furnaces unremunerative even in periods of great activity.

Pennsylvania shows a decrease of 45 furnaces from 1880 to 1890, and during the same period the total number of furnaces in Ohio has decreased by 32. These figures, however, merely exhibit the net decrease in the number of furnaces, as many large bituminous coal and coke furnaces have been erected during this period in these as well as in other States to take the place of small stacks abandoned. Since 1880 there have been 282 furnaces abandoned in the United States, owing either to unfavorable location or to give place to larger and more modern plants, while during the same period 163 new furnaces have been built, in addition to a large number of plants that have been remodeled and new machinery added.

At the close of the census year 1890 the total number of blast-furnaces which were active, or likely to be some day active, was 562, of which 169 were anthracite or anthracite and coke furnaces, 253 coke and bituminous coal furnaces, and 140 charcoal furnaces. Of the total number of furnaces at the close of 1880 there were 229 anthracite or anthracite and coke furnaces, 195 coke and bituminous coal furnaces, and 257 charcoal furnaces. In the decade from 1880 to 1890 there is seen to have been a decrease of 60 in the number of anthracite or anthracite and coke furnaces, a decrease of 117 in the number of charcoal furnaces, and an increase of 58 in the number of coke and bituminous coal furnaces.

Of the 562 completed furnaces at the close of the census year 1890 there were 338 in blast, of which 110 were anthracite or anthracite and coke furnaces, 165 coke and bituminous coal furnaces, and 63 charcoal furnaces. The number of furnaces building at the date mentioned was 39, of which 9 were in Virginia, 7 in Alabama, 5 in Pennsylvania, 4 in Illinois, 3 each in Kentucky, Tennessee and Michigan, 2 in Maryland, and 1 each in Georgia, Ohio and Wisconsin.

The total number of Bessemer steel plants at the close of 1889 was 53, with 108 converters, including 6 Clapp-Griffiths plants with 11 converters and 7 Robert-Bessemer plants with 11 converters. Forty works, having 84 converters, including 5 Clapp-Griffiths plants with 9 converters and 6 Robert-Bessemer plants with 10 converters, were employed during 1889 in the production of Bessemer steel.

Pennsylvania made 60 per cent. of all the Bessemer steel ingots produced in 1889, against over 56 per cent. in 1888, 53 per cent. in 1887, 59 per cent. in 1886, and 65 per cent. in 1885. Illinois made 22 per cent. in both 1888 and 1889, 26 per cent. in 1887; 21 per cent. in 1886, and 22 per cent. in 1885.

The total number of completed open-hearth steel works in the United States at the close of 1889 was 56, two more than at the close of 1888.

In a table the marvellous growth of the Bessemer steel industry is shown for sixteen years: The total output of 1874, viz., 191,933 net tons having increased to 3,281,829 tons in 1889, and a similar development of the open-hearth steel industry for the same time is exhibited by the product of 7000 net tons in 1874 increasing to 419,488 net tons in 1889.

We note that 331,283 net tons of plate and sheet steel and 471,193 net tons of plate and sheet iron were produced in 1889.

DR. W. M. SWEET has also issued a bulletin on steel production, from which the following excerpts have been taken to show the growth of this industry:

"The table given below shows the production of various kinds of steel in the form of ingots or direct castings in the census years of 1880 and 1890:

Kinds of Steel. (Ingots or Direct Castings.)	Tons of 2000 Pounds.	
	Year ended May 31, 1880.	Year ended June 30, 1890.
Bessemer steel	985,208	3,788,572
Open-hearth steel	84,302	504,351
Crucible steel	76,201	85,536
Clapp-Griffith steel	....	83,963
Robert-Bessemer steel	....	4,504
Total	1,145,711	4,466,926

"During 1880 fourteen States contained steel-making establishments, and steel was produced in that year in each of these States except Rhode Island and Maryland. In 1890 steel works were located in nineteen States, and steel was made in that year in each of these States except Kentucky, Missouri and Virginia.

"Pennsylvania continues to occupy the position of the leading producer of steel in the United States, producing 57 per cent. of the total production in 1880, and 62 per cent. in 1890. Illinois was second in rank in both years, and Ohio was third. From 1880 to 1890 the increase in production in Pennsylvania was 324 per cent., in Illinois 241 per cent., and in Ohio 314 per cent. Since 1880 the manufacture of steel has been abandoned in two States, viz., Rhode Island and Vermont, and seven States have engaged in its production, viz., Alabama, California, Colorado, Indiana, Michigan, Virginia and West Virginia.

"The increase in the number of establishments producing Bessemer steel has been the result almost entirely of the demand for steel in forms other than rails. All of the eleven Bessemer steel plants that were completed in 1880 had been built to manufacture steel for rails, many of them being added to previously existing iron rail mills. Of the fifty-three Bessemer steel plants at the close of 1890, only fourteen made steel rails during that year, and of the total quantity of rails produced over 90 per cent. was made by ten of these works. Thus, while the production of steel rails has nearly trebled in amount since 1880, the number of establishments engaged in their manufacture has shown but little change in the ten years, although many of these works have greatly increased in size and efficiency. The competition in the manufacture of Bessemer steel rails has compelled many of the rail mills to convert a large part of the steel produced by them into forms other than rails, the production of rails to any considerable extent at the present time being possible only in works favorably located for the supply of cheap raw materials and operated under the latest and most improved methods of manufacture.

"The increased quantity of Bessemer steel manufactured in these miscellaneous forms is approximately shown by a comparison of the ingots and rails produced, over 75 per cent. of the ingots made in 1880 being converted into rails, while in 1890 the percentage of rails made to ingots produced was only 53 per cent.

"While the basic process is applicable to either the Bessemer or open-hearth process, its use in this country in connection with the open-hearth furnace is most promising of successful results, and the indications are that the growth of the basic steel industry of the United States will be largely in this direction. The total production of basic steel in the United States during 1890 amounted to 62,173 tons of 2000 pounds, nearly all of which was made by the basic open-hearth method, a small part being produced by the duplex process, a combination of the Bessemer and open-hearth methods."

Concerning the iron trade of the world, Mr. Swank says: "In the year 1889 the whole world was at peace, and the arts of peace flourished as they never flourished before. The world's production of iron and steel was very much larger in that year than in any preceding year. The United States virtually supplied its own increased iron and steel wants, except for Spiegeleisen, ferro-manganese, and tin plates, but the extraordinary demand for iron and steel from other countries could only be met by European manufacturers. The year 1889 was therefore one of the most active and one of the most prosperous in the history of the European iron trade, notwithstanding the reduced shipments of iron and steel which were made to the United States. Every European iron and steel making country shared in the year's prosperity. Prices were high all through the year, but highest at its close."

COMPARATIVE ANNUAL PRODUCTION OF PIG-IRON.

Years.	Great Britain. Gross tons.	United States. Gross tons.
1882	8,586,680	4,623,323
1883	8,529,300	4,595,510
1884	7,811,727	4,097,868
1885	7,415,469	4,044,526
1886	7,009,754	5,683,329
1887	7,559,518	6,417,148
1888	7,998,969	6,489,738
1889	8,245,336	7,603,642

The preceding table gives the production of pig-iron in Great Britain in the last eight years, compared with the production in the United States. The British maximum was attained in the year which begins the table.

The United States is increasing its production of pig-iron so rapidly that it is evidently destined to become at an early day the

COMPARATIVE PRODUCTION OF IRON ORES IN VARIOUS COUNTRIES.

Countries.	Iron Ore.	
	Years.	Tons.
Great Britain	1888	14,590,713
United States	1889	14,096,427
Germany and Luxemburg	1888	10,664,800
France	1889	2,500,000
Belgium	1888	213,000
Austria and Hungary	1888	2,200,000
Russia	1887	1,334,699
Sweden	1889	985,904
Spain	1889	4,500,000
Italy	1887	230,575
Other Countries	1889	2,000,000
Total	....	53,316,118
Percentage of the United States		26.4

leading pig-iron producing country in the world, possibly reaching this distinction in 1890.

The number of blast-furnaces existing in Great Britain at the close of 1889 was 813, of which 461 were then in blast, against 429 on March 31st, 434 on June 30th, and 445 on September 30th.

In the succeeding paragraph reference is made to the larger importations of tin plates from Great Britain into the United States. We trust that the policy lately adopted by our government of encouraging temporarily this valuable industry will rapidly reduce the imports of this commodity and place the United States so well towards the front rank as a producer of tin plate, that the protection offered will be permanent.

The exports of tin plates from Great Britain in 1889 were the largest ever recorded, the total exports amounting to 430,623 tons, of which the United States took 336,692 tons, or over three-fourths

WORLD'S PRODUCTION OF PIG-IRON AND STEEL.

Countries.	Pig Iron.		Steel.	
	Years.	Tons.	Years.	Tons.
Great Britain..................	1889	8,245,336	1889	3,669,862
United States..................	1889	7,603,642	1889	3,385,732
Germany and Luxemburg.	1889	4,387,504	1888	1,862,000
France..........................	1889	1,722,480	1889	529,021
Belgium..........................	1889	847,000	1889	248,000
Austria and Hungary	1888	761,606	1888	355,038
Russia	1887	532,649	1887	222,025
Sweden	1889	420,665	1889	137,821
Spain	1888	200,000	1887	24,500
Italy..........................	1887	12,265	1887	73,262
Other Countries..............	1889	100,000	1889	30,000
Total	....	24,833,147	....	10,537,261
Percentage of the United States.....		30.6	....	32.1

of all, for which the people of the United States paid the Welsh tin-plate manufacturers £4,674,455, or nearly $23,000,000. Germany has entered upon the manufacture of tin plates. The production of tin plates in Germany in 1887 was 16,720 metric tons, basic steel being almost exclusively used.

The preceding tables show our production of iron-ore, pig-iron and steel in 1889 in comparison with their production by other countries in that year or in the most recent years for which official statistics or data for a careful estimate are available. Gross tons of 2240 pounds are used in giving the statistics of Great Britain and the United States, and metric tons of 2204 pounds are used for all the continental countries of Europe.

The world's production of pig-iron increased from 14,117,902 tons in 1878 to 24,869,534 tons in 1889, or 76 per cent. while the world's production of steel increased in the same period from 3,021,-093 tons to 10,513,977 tons, or 248 per cent. This is wonderful progress. The figures we give are most significant, however, in showing how rapidly the use of steel has grown in favor, notwithstanding the increased use of manufactured iron.

Irondale Furnace, Washington.

THE blast-furnace of the Puget Sound Iron Company has the distinction of being located further north than any other iron producing plant in the United States. The original furnace which was erected in 1882 was dismantled, and in 1884 a new and larger one was placed in the immediate vicinity, but on higher ground. The furnace plant is located at the head of Port Townsend Bay, an estuary of Puget Sound. The settlement about the plant which is known as Irondale, is about 7 miles from Port Townsend. The plant consists of an iron shell furnace stack supported upon iron columns (located in a frame casting house), an engine house, a 60-pipe Player hot-blast stove, 2 batteries of boilers, a charging house, a hoist tower, a charcoal house and 20 charcoal kilns.

The furnace stack is lined to a bosh diameter of 11 feet with crucible 5 feet in diameter; it has steep boshes, and is fitted with a bell, hopper and seal, the bell being 3 feet 6 inches in diameter, for a tunnel head 7 feet 9 inches diameter. The furnace is 50 feet in height, and the hoist tower which extends from wharf level has a vertical lift of 85 feet. Blast is delivered into the furnace at a temperature of 800° to 850° F., through five 3¾-inch tuyeres at a pressure approximating 4 lbs. It is supplied by a vertical Weimer engine, steam cylinder 30 inches in diameter, air cylinder 72 inches in diameter, stroke 4 feet. There is also in the engine house two other engines which are not in use.

The location of this furnace was originally decided upon by the existence of beds of brown hematite-ore in the vicinity, and by the abundance of wood from which to make charcoal. This ore is a bog formation, which when exploited failed to sustain in quantity the expectations of the projectors. In addition the ore encloses large numbers of pebbles, which reduce the available amount of metallic iron to about 30 per cent. The lean character of the ore and the possible exhaustion of convenient beds of it, led the company to look elsewhere, and some 2500 acres of mineral land were purchased on Texada Island in the Gulf of Georgia, British Columbia, located about 100 miles north of the furnace. The ore from Texada Island is a hard dense magnetite, in which sulphur occurs as pyrites, and the practice is to roast this ore in heaps before charging it in the blast-furnace. This magnetite-ore yields over 60 per cent. of iron in the furnace, and through the courtesy of Mr. George Froescher (Manager of the Puget Sound Iron Company) who furnished most of the facts and figures gleaned during our visit to the Irondale Furnace, we are indebted for the following analyses. The Texada ore forms the base of the present ore supply of the Irondale Furnace, only sufficient of the local ore being employed for the production of cinder. Mr. Froescher informs us that he has on occasions used the Texada ore alone, adding brick-bats to the charge on account of the ore working so dry. This ore coming from British Columbia has to pay a duty of 75 cents per ton.

The limestone is obtained from the San Juan Islands in Puget Sound 30 miles distant. Charcoal is produced from the native fir which abounds in the vicinity, and which seems to reach its best stage of development in Puget Sound country. This wood is felled and cut by Chinamen with saws, so that full cords are obtained. One dollar is paid per cord for felling, cutting and splitting.

The wood is delivered to the furnace mainly on scows although some of it is hauled in by teams. The charcoal-kilns are located at wharf level and are of the conical form of the following dimensions: Diameter at bottom 30 feet, height 30 feet, radius from which the sides are formed 37 feet 6 inches. The lower portion of the walls for 6 feet is made of one and a half bricks, the upper portion is one brick thick, laid in a mixture of lime with cement mortar. In addition to the upper charging door and lower charging door there is an explosion door at the top, and a band of 3 x $\frac{1}{4}$-inch iron is placed around the base of the 9-inch wall. Mr. Froescher states that these kilns each hold 75 cords of wood and produce on average 4200 bushels of charcoal (2815 cubic inches per bushel), equivalent to 56 bushels per cord. This apparently enormous yield is owing to the large size and straight grain of the wood and the sticks being sawed and split, thus permitting them being closely ranked for measurement and packed tightly in the kiln. The coal is, however, light, a bushel of 2815 inches weighing 16½ to 17 pounds, and about 160 bushels are required to produce a ton of pig-iron. The ore and limestone is delivered from vessels into "bunkers" at the wharf-level, and from these discharged into trollies on tramways leading to the roasting piles and crusher. From the crusher a bucket elevator delivers ore and flux into appropriate bins, from which they are shovelled into charging barrows which are run on to the large double hoisting cages and elevated to the top of the furnace. As the furnace stack stands 35 feet above the wharf, the total lift, as above stated, is about 85 feet. The pig-iron made is used in the ten or twelve foundries on Puget Sound, and a considerable market is also found for it in San Francisco; it is used chiefly for car wheels and other special castings, as the foreign pig-iron for ordinary foundry work can be delivered along the Pacific coast at from $25 to $28 per ton. It is not improbable, however, that this furnace or others which are projected, will obtain their supply of ore from mines in the State of Washington, for there is an abundance of iron-ore convenient to Puget Sound, which is only waiting for transportation facilities to utilize it. An experiment was made at the Irondale Furnace in which coke produced from Washington coal gave satisfactory results. In fifteen months the Irondale Furnace produced 13,000 tons of pig-iron, mostly of foundry grades.

Analyses of iron-ores used at the Irondale Furnace:

Bog-ore from Chimacum Valley, Jefferson County, Washington.

Moisture,	16.40		
Combined water,	12.80		
Silica,	8.02		
Sesquioxide of iron,	52.21	Iron,	36.55
Sesquioxide of manganese,	3.02	Manganese,	2.11
Alumina,	3.33		
Lime,	1.01		
Magnesia,	0.15		
Phosphoric acid,	2.605	Phosphorus,	1.137
Sulphuric acid,	0.520	Sulphur,	0.21
Total,	100.065		

Mr. Froescher considers this above the average yield of iron, and quotes the following as more closely approaching furnace results:

Iron,	28.48
Phosphorus,	0.711

Magnetite from Texada Island, Gulf of Georgia, B. C.

Silica,	5.09		
Sesquioxide of iron,	55.30	Iron,	62.63
Protoxide of iron,	27.56		
Bisulphide of iron,	5.31	Sulphur,	2.837
Alumina,	2.79		
Lime,	3.31		
Magnesia,	0.21		
Phosphoric acid,	0.043	Phosphorus,	0.019
Titanic acid,	0.32		
Total,	99.933		
Phosphorus in 100 parts of iron,	0.03		

Another determination of this same ore gave:

Silicious matter,	5.35
Sulphur,	0.93
Phosphorus,	0.02
Metallic iron,	64.40
Phosphorus in 100 parts of iron,	0.031

Blast-Furnaces as Town Attractions.

A PROMINENT politician has lately attracted attention by prophesying a financial crisis, as the result of mortgaged lands in the northwestern States, and this prophesy might have specified as another and greater source of danger, the craze for town developments which is so prevalent.

The JOURNAL does not propose to discuss the relative values of real estate in various sections of the country, nor offer any opinion as to what towns have the best promises of future greatness, but it does intend to caution its readers against a whirlpool of excitement which may ultimately engulf the earnings of many years, bring financial ruin upon industrious citizens, and encourage distrust in the production and manufacture of iron as a business. It is the last-named reason which induces us to discuss the present excitement existing in some portions of our country.

There is no question as to the value which an active blast-furnace is to a community, nor as to the local development which a rolling-mill or other iron or steel manufacturing industry encourages, but the expectation that a city is to develop into existence, on a particular site selected by the promoters, because of the establishment of blast-furnaces or iron-works, has but slight foundation, and there is still less basis for the valuation of land in the midst of a wood on a prospective street or avenue of an embryo city at prices closely approximating and sometimes exceeding those prevailing in our large trade centres.

The existence of a deposit of iron-ore is not in itself encouragement sufficient for the erection of a blast-furnace, and the fact that all the raw materials requisite for the production of pig-iron occur close together, may, by reason of the quantity and quality of these, or of market considerations, demonstrate that the erection of blast-furnaces or the establishment of kindred industries would be inadvisable at present.

The location of iron industries in the heart of populous residence districts has not been generally approved, the policy being rather to have small individual communities centre about iron-works, or else locate them in suburbs of large towns and cities and away from the residence or mercantile sections.

Probably not less than one hundred cities have been lately laid out, or are now promised by liberal advertisement, all based upon the erection of iron-works. Invited by display advertisement and elaborate prospectus, accompanied by announcement of other attractions, crowds often of great proportions assemble at auction sales of lots, and at some of these sales property has been disposed of for over half a million dollars, which a few weeks before would not command 1 per cent. of the price received, and in a majority of instances the basis of the magic increase in valuation was the promised iron industries. It is hardly probable that any one not blinded by enthusiasm will claim that the prices realized represent intrinsic values, for were such the case the boomers and speculators who frequent these sales would not unload so quickly, nor will any disinterested person admit that prospective building lots, located upon swaths cut through forests and labelled with street and avenue names, can compare in value with those fronting on thoroughfares which are paved, sewered, supplied with water and light facilities, and which are adjacent to other properties upon which are substantial improvements. But to-day, in numerous localities, the future is thus discounted to an unprecedented extent, lots located in the portion of the proposed city which is presumed to be the business district, bringing prices in advance of what similar lots command in large communities whose industries have been long established, and residence lots are scheduled at figures which are beyond the ability of the average mechanic to pay rent for or interest on at prevailing rates of wages. We confess that we fail to see in high land values any advantages to manufacturing, and we know it is to the detriment of the iron trade.

Some of those who served as officers of the Confederate army preserve as relics boots which cost them $250 per pair, or suits of clothing for which $1000 or more was paid in Confederate money, and in future years a resident in one of these boom cities may be able to astonish visitors by the statement that the lot on which his home stands, or possibly the field which he is plowing, sold in 1889–1890 for what will then be recognized as a fabulous price.

We do not desire to be understood as condemning proper effort to attract notice to the merits or advantages of a locality; on the contrary, we believe many communities only need to appreciate their resources and advantages and make these known to others to effect advancement which will be startling. We are also sufficiently conversant with our iron industry to appreciate the help it is to a community, but we do not believe that a blast-furnace located primarily to boom a town site will be as successful as one whose situation is decided upon because of the available supplies of raw material and convenience to a market for product, nor do we believe that scattering blast-furnaces or rolling-mills in certain localities as seed will necessarily germinate into crops of magnificent cities.

That disaster will follow many of these schemes seems certain, for when the final payments to be made fall due, after the boom has passed, when transfers of property executed in haste develop faulty phraseology, or vitiate titles, there will be ample business for lawyers if the properties are then worth a contest, and if not, then the loss will fall on the last one in possession. Although some of these speculations have netted the promoters sufficient to pay the expenses of constructing blast-furnaces or other iron industries, as well as a handsome profit, the stability of these works for the future is not always provided for. The failure of the iron-works will cripple the entire community whose establishment it encouraged, and the collapse of the *enterprise* will be laid to the discredit of iron industries, and this discredit may result in the serious financial annoyance of works whose location, equipment and management entitle them to confidence.

Cast-Iron Car Wheels.

IN an article upon "Cast-Iron Car-Wheels," which appeared in the JOURNAL, vol. viii., pages 153 to 170, we gave a synopsis of a very interesting discussion which ran through several sessions of the New York Railroad Club. Among the prominent features of this discussion was a reference to contracting chills, the merits of various forms having been discussed, and the conduct of the metal from the time it was poured into the mold until the wheel was removed, was explained by various parties from different stand-points of practice or experience.

There has been a great deal said about the mileage of wheels which were made by the various car-wheel foundries, and the merit of certain mixtures or specific methods of producing wheels has been largely proclaimed. On the other hand, there has been a very severe duty-test assigned to wheels by the users of them, a test none too hard, in view of the fact that so many lives and so much property is put in jeopardy by every imperfect car-wheel, but there has, no doubt, been some hardship in the interpretation of responsibility which called out, at the meeting of the Manufacturers of Chilled Car-Wheels at New York, last fall, a protest, on account of the wheel-makers having no control over the condition of railroad service, which took the form of the following resolution:

"*Resolved*, 1. That, in all mileage or time guarantees, the wheel-maker ought to be held responsible only for wheels which fail through faults of material or workmanship.

"2. That, when wheels are taken out of service, on account of sharp flanges, flat spots, comby or shelled out treads, or for cracked brackets or plates, and it is found, on breaking up the wheels, that the depth and character of the chill, and the strength and character of the metal in the plates, are up to the standard specifications adopted by the joint conference committee of the Railway Master-Mechanics', the Master Car-Builders' and the Wheel-Makers' Associations, it shall be considered that the failure is due to the service, and not to the quality, of the wheels, and that the wheel-maker ought not to be called upon in such cases to pay for or replace any such wheels."

In the matter of tests of wheels, we present a statement, issued by Messrs. A. Whitney & Sons, of Philadelphia, giving the results of inspections and tests of 10,000 wheels from their foundry, made between March, 1889, and March, 1890, by the R. W. Hunt & Co. Bureau of Chicago, and we give the record of this inspection to show the perfection to which the casting of car-wheels is now being carried.

Each wheel was measured around the tread by a brass tape, the record showing that 4325 were all of exactly the same circumference; 4365 were all of exactly the same circumference, but one-eighth inch less than the 4325; 881 were all of exactly the same circumference, but one-eighth inch more than the 4325; 429 varied one-eighth of an inch, either way, from the above.

Out of the 10,000, the extreme variation in 8690 was only $\frac{1}{8}$ of an inch in circumference, or $\frac{1}{24}$ of an inch in diameter, and in 881 more it was only $\frac{1}{4}$ of an inch in circumference, or $\frac{1}{12}$ of an inch in diameter.

Tested by a true ring resting on the cone of the wheel, not one was found with a variation of more than $\frac{1}{32}$ of an inch from the ring at any point.

To test the strength, the specifications required that one wheel, out of every 100 ready for shipment, should be broken by a drop of 140 pounds weight falling twelve feet, and that it should stand five blows, without breaking into two or more pieces. For the 10,000 wheels shipped, 105 were thus broken, with the following results:

To start the first crack, the average number of blows was $13\frac{26}{105}$.

To start the first crack, the weakest required 5 blows, and the strongest 58.

To break the wheel in two, the average number of blows was $49\frac{56}{105}$.

To break the wheel in two, the weakest required 9 blows, and the strongest 118.

(Two wheels broke at 9 blows, fifteen between 10 and 20, four between 20 and 30, thirty-four between 30 and 50, and fifty between 50 and 118 each.)

The Messrs. Whitney say, the average depth of chill, at the root of the flange in the 105 wheels broken, was $\frac{7}{16}$ of an inch, and the depth did not vary, in any case, more than $\frac{1}{8}$ of an inch around the wheel.

Not a single wheel was rejected from the whole number offered for inspection because of any chill-cracks, blow-holes, or other imperfections of workmanship or material and the metal, in all the wheels broken up, was found to be of a good, soft, open, gray in the plate and free from slag or dirt. The tread of every wheel was practically as smooth as if cast in an ordinary solid chill, and not a single one needed to be ground, or in any other way made smooth or true.

Messrs. A. Whitney & Sons claim that these results are largely due to the fact that the wheels are cast in their contracting chills which are so constructed that, when the wheels are poured, the heat of the molten metal causes the inner surface to *contract* both in diameter and circumference. This is accomplished by making the chill to consist of two rings, separated by air-spaces, and dividing the inner ring, in the process of casting, into more than one hundred segments by the use of asbestos cores. Each kerf is thus not much more than $\frac{2}{100}$ of an inch wide, but, together, they permit a contraction of two inches in the circumference of the inner ring. The ridges made by them are, it is claimed, so small that the treads of the wheels are practically as smooth as if cast in solid chills, while the chill on the tread is deep, uniform and hard, without hardening the plates or hubs. Messrs. Whitney also claim that their *mixture* of irons is made up exclusively of the best *charcoal brands*, each iron being carefully and thoroughly tested before it is accepted, and is admitted to the mixture only when it possesses superior chilling properties or strength.

There are other forms of contracting-chills, for which advantages are claimed, equal to those above enumerated, and there may have been other tests which show results comparing favorably with what has here been given, but we have not reports of such tests to offer our readers. It will, however, gratify us to be able, in a future issue, to add to the data herein given, by publishing records of tests of car-wheels, cast by various foundries, produced in a special manner, particularly where the report of tests is made by a disinterested party of such high repute as the R. W. Hunt Testing Bureau.

In reference to the contracting-chill, we take this opportunity of quoting from a communication to the National *Car and Locomotive Builder*, in which Mr. John R. Whitney discussed the behavior of iron in solidifying.

For the 'whole' casting, the process of solidification is a gradual one, and with some irons it is much more gradual than with others. But it does not follow from this that a casting 'will, from the moment the mold is filled or poured till it has reached, as a casting, its lowest point of temperature, be all the time constantly shrinking and contracting so as to occupy less apace.' If it did, in every casting, and especially in every one of any considerable size, the centre ought to be the most intensely compressed part, as it is the last to reach the 'lowest point of temperature,' and, therefore, should show the finest and closest grain. But we do not find it so. The centre always shows the largest and most open grain. If, however, as each successive portion becomes solid, its contraction is resisted, and prevented by expansion of each successive portion in becoming solid, the open-grained centre is easily and satisfactorily explained.

"Several years ago, I prepared two molds of exactly the same size in every particular. One mold was filled with iron at a white heat, directly from the cupola. The ladle was then allowed to stand until the rest of the iron had cooled down so as to be only just fluid enough to run freely, when the second mold was filled. Both were left undisturbed in the sand until they were perfectly cold, then cleaned and carefully measured. They were found to be of exactly the same dimensions in every particular. The shrinkage of each from the size of the pattern was exactly $\frac{25}{100}$ of an inch in 12 inches.

"On another occasion, an ordinary pig, about 8 feet long, was molded in open sand so that one end of the mold was closed by a solid brick wall, and the other end by a fire-brick, free to move in the sand. Against this fire-brick a bar of iron, 6 feet long and 1½ inches square, was laid, one end being pivoted to the ground and the other resting on a fire-brick. The fire-brick connection with the mold was at about 2 feet from the pivoted end.

This apparatus was prepared impromptu, and was exceedingly crude. I expected no satisfactory results. On pouring the pig, however, every one was surprised, after a few minutes, to see the bar of iron gradually move on the supporting fire-brick until it lay ¼ of an inch from its original position.

"On a more recent occasion, the following experiment was made with an apparatus more carefully prepared: A pattern, 4 feet long, 3⅞ inches deep and 2¾ inches wide, was molded in open sand, one end of the mold being closed by fire-brick and and the other end by a piece of gas-carbon which was suitably connected with a small battery and galvanometer. The fire-brick rested at one end against a block of iron weighing about half a ton. The gas-carbon block was carefully secured in the sand, so that the weight of iron in the mold should not be sufficient to move it, and the stand, bearing an arm on which a pointer was delicately pivoted, was then adjusted so that the needle should press against the gas-carbon, and the pointer stand at zero on the scale. The long arm of the pointer was 24 inches, and the short one 6 inches long, or, as 1 to 4. The scale was graduated to $\frac{1}{16}$ of an inch.

"The mold was filled with very fluid hot iron in seventeen seconds, and then the following results were carefully noted:

For more than 1 min. after the mold was filled, pointer stood at zero.						
At 1 min. 30 sec.	"	"	"	"	moved 1.16 inch.	
" 1 " 50 "	"	"	"		had moved ⅛	"
" 3 " 10 "	"	"	"		" ¼	"
" 5 " 20 "	"	"	"		" ⅜	"
" 8 " 5 "	"	"	"		" 7-16	"
" 11 " 30 "	"	"	"		" 15-32	"
" 12 " 5 "	"	"	"		" ½	"

"From that time, the pointer stood perfectly still at ·½ inch until twenty-five minutes and fifteen seconds after mold was filled, when the galvanometer showed that contact with the gas-carbon was broken and shrinkage had begun.

"Long before these experiments were instituted, the fact that iron follows essentially the same law as water in solidifying, was well known and published. I need cite only two authorities. Professor Edward Turner, in his *Elements of Chemistry*, published in Philadelphia in 1835, by Desilver, Thomas & Co., says, p. 20: 'Water is not the only liquid which expands under reduction of temperature, as the same effect has been observed in a few others which assume a highly crystalline structure in becoming solid; fused iron, antimony, zinc and bismuth are examples of it. Professor Thomas Graham, also, in his *Elements of Chemistry*, published in Philadelphia, in 1843, by Lee & Blanchard, says, p. 385: 'Iron expands in becoming solid and, therefore, takes the impression of a mold with exactness.'

"As the observation of this law was the basis upon which my experiments leading to the successful development of the contracting chill for cast-iron car-wheels, was based, I am persuaded that it will lead to many other practical results of great importance.

"JOHN R. WHITNEY."

Swedish Mineral Statistics.

THROUGH the courtesy of Prof. Rich. Akerman, of Stockholm. Sweden, honorary member of the Association, we are supplied with the latest official statistics of the iron and steel industries of Sweden.

From the following table it will be seen that the 1888 output of iron-ore was increased 26,364 tons in 1889, while the production of charcoal pig-iron decreased nearly 8 per cent. This would seem to prove either a decrease in the yield of iron in the ore or that there has been a large increase in the exports of iron-ore. In nearly all of the other kinds of manufactured iron and steel there has been a marked increase:

PRODUCTION OF IRON AND STEEL IN SWEDEN, METRIC TONS.

	1882.	1883.	1884.	1885.	1886.	1887.	1888.	1889.
Iron-ore	892,863	885,124	909,553	873,362	872,479	903,186	959,540	985,904
Charcoal pig-iron	398,945	422,627	430,534	464,737	442,457	456,625	457,052	420,665
Bar iron and rods made	259,462	255,853	264,944	257,369	237,130	255,383	253,090	274,734
Bessemer iron and st'l made	47,358	50,878	53,123	52,021	54,121	68,199	68,620	80,324
Martin iron and steel made.	13,405	16,800	19,354	26,743	22,361	41,898	44,712	55,487
Other kinds of steel made	1,430	1,827	1,764	1,786	1,749	1,468	1,205	2,010
Plates made	15,805	17,439	17,534	16,494	13,579	12,394	19,701	27,389
Nails made	8,143	8,197	9,720	10,577	10,289	10,239	10,683	12,072
Number of blast-furnaces in operation	185	191	178	179	164	164	162	150
Total time for all furnaces in blast, days	40,157	41,229	40,361	42,460	39,777	40,582	39,840	35,859
Average time for furnaces in blasts, days	217	216	227	237	242.5	247.5	246	239
Average production per furnace and day	9.93	10.25	10.67	10.95	11.12	11.23	11.47	11.73

The number of active blast-furnaces has fallen off, and the tótal and average time for all the furnaces is also less, but the average daily production per furnace has increased, showing a gradual and constant advance in blast-furnace practice.

A Congress of Metallurgists.

THE autumn of 1890 should be memorable to all interested in iron metallurgy on account of the congress of the three leading associations of the world which devote attention to iron mining, iron and steel metallurgy, or kindred industries. Some years ago the American Institute of Mining Engineers extended an invitation for the Iron and Steel Institute of Great Britain to hold a meeting in the United States. Several prospective dates were named, but disappointment followed in each case until the present year, when the invitation was formally accepted, and arrangements for a meeting in the United States were perfected. By similar invitation the Verein Deutscher Eisenhuttenleute, in whose membership is included the leading iron masters, metallurgists and engineers of the German Empire, also came to the United States at the same time.

In the latter part of September the American Institute of Mining Engineers held its meeting in New York city, and the sessions of this association were followed by those of the Iron and Steel Institute. After five days devoted to business and receptions, the representatives of the three organizations started on a tour of inspection of some of the resources and industries of the United States. The points visited by the entire excursion, which numbered over six hundred, were Philadelphia, Mount Gretna, the Cornwall Ore Banks, Lebanon, Altoona, Johnstown, Pittsburgh and the oil regions, coke region and industries tributary thereto in Pennsylvania. The excursion then proceeded to Chicago. It is hardly necessary to say that the two hundred and seventy-five gentlemen and ladies representing the Iron and Steel Institute, and the one hundred and seventy-five who wore badges of the Verein Deutscher Eisenhuttenleute, received, *en route*, a continuous ovation and many public and private courtesies. At Chicago the party divided, one large train of eleven Wagner vestibule cars going north, and two trains of Pullman vestibule cars proceeding southward. The northern party devoted a week to inspecting the iron and copper mines of upper Michigan; then passed, *via* the Sault ste Marie, into Canada, visiting copper and nickel deposits in the vicinity of Sudbury. After a day spent at Niagara, the northern contingent proceeded to Washington, where they met the southern party. The southern party travelled directly from Chicago to Birmingham, Ala., devoting a week to the examination of the blast-furnaces, iron-ore mines, etc., of Alabama, eastern Kentucky and Tennessee, and southwestern Virginia, meeting the northern party at Washington, where a reception was given to the reunited delegations by the President of the United States. After a visit to the new blast-furnace plant at Baltimore, the party again separated, the majority going to New York and taking passage homeward; a considerable delegation, however, returned to Niagara, and were there taken in charge by the Canadian Government, devoting a week to visits to the prominent cities of the Dominion, returning *via* Boston to New York.

The trips, as laid out, were not calculated to exhibit to our visitors the agricultural or commercial developments of the country, nor to visit populous cities. The programme was so arranged as to exhibit the resources most in harmony with the interests of the different organizations which were guests of the American engineers; hence, in the northern excursion, none of the large cities, except Chicago, were visited; and in the southern trip, stops were made at but two or three of the more prominent cities. Ample opportunity, however, was given to appreciate the mineral resources and the metallurgical developments for utilizing them, and our European friends undoubtedly carried away with them impressions which will be of benefit to themselves and possibly of service to this country.

The personnel of the visiting delegations was of a high grade; much more so, in fact, than would generally be expected in an excursion of organizations covering so large a territory and demanding so much time. The most prominent men in the English delegation, as well as in the German delegation, were the honorary members of the United States Association of Charcoal Iron Workers. These were Sir Lowthian Bell, of England, and Dr. Herman Wedding, of Prussia. It was a pleasure for us to renew the acquaintance with these gentlemen, formed during their former visits to the United States. Among both delegations were members whose names are recognized as authorities, not only in their own country, but all over the metallurgical world, and we only shrink from mentioning these because we scarcely know where to draw the line, there being so many members of importance and merit.

A meeting such as that referred to, where good fellowship born of kindred interests was the only cause to bring such a delegation together, cannot help but be fruitful of future good, and the result of such assemblages, wherein personal acquaintance and individual attachments are formed, will do more to cement friendship of countries than any treaty which can be written.

The invitations were sent because the American Institute of Mining Engineers recognized the service which the kindred organizations of Great Britain and Germany had been to the development of their own and American industries, and these invitations were accepted in the same spirit. The American manufacturers of iron and steel, and those interested with them, contributed liberally to a fund placed in the hands of a general committee, of which Mr. Andrew Carnegie was chairman and Mr. Charles Kirchhoff, Jr., secretary. In addition, the local engineers and manufacturers in the various sections of the country covered by the excursion made generous provision for the entertainment of their guests.

It was our privilege to be associated with the Central Committee as a representative of the Association, and also individually with some of the Local Committees, and to participate in a portion of the excursion as a representative of these committees. It is with pleasure that we bear testimony to the delightful manner in which the efforts of the Americans were appreciated by their guests, and to the expressions of good-will and fraternal feeling which were given throughout the journey.

Our observation convinces us that the visiting engineers were most impressed during the trip by the labor-saving devices employed in this country, by the liberal use of electricity as an illuminating agent, and by the energy and vim which characterizes the management of our large mines, blast-furnaces and other industries. They return to their own country with a practical knowledge of some of the resources and possibilities of the sections of the country through which they passed, having acquired a knowledge of American methods and practice. Our American engineers learned fully as much from our visiting friends as we communicated to them; and some of the guests, whose names had been associated with the development and improvements in metallurgy, were treated with such courtesy and attention as to border on embarrassment, as they expressed it.

The results of this visit will not be immediately apparent. Undoubtedly, the investigations made by individuals will encourage some foreign capital to seek investment in the United States, and, on the other hand, a personal inspection of some of the unfortunate local methods pursued in following a practice which we have Americanized as "booming an enterprise," will discourage the investment of money in those directions, and it is better for the country that this should be the case. The visit will also result in the permanent location of some who came to this country as guests, and others have expressed an intention of having members of their family locate in the United States.

The freedom with which our industries were opened for inspection for our visitors was a topic much commented upon, and we rejoice that such was the case. We shall not be surprised if some have carried home valuable ideas which will assist them in competing with our own manufactures, nor do we lose sight of the fact that impressions gained in this country may encourage the application of labor-saving appliances to a greater extent than is now the case. But in any retrospect we take of the Congress of Metallurgists, we cannot but feel that, ultimately, America will be the gainer from the visit made ; and we rejoice that so large an excursion was permitted to cover so many miles of travel with no serious drawbacks, and with no general sorrow, or no individual disaster to mar the delightful recollections which follow it.

We shall present *résumés* of some of the papers and discussions which formed a portion of the meetings of the three Associations, in this and subsequent issues of the JOURNAL.

SINCE the last issue of the JOURNAL the Association has lost a member, who showed a lively interest in its welfare, in the person of Dr. Frank King, of Van Buren Furnace, Shenandoah county, Va.

Dr. King was well known to most of the members of the Association, having served as Vice-President during the year 1882, and as a member of the Board of Managers for four years, 1885 to 1888 (inclusive). His intelligence and genial social characteristics endeared him to many, who will mourn his loss as a friend.

The Concentration of Iron-Ores, No. 6.

THE interest which has been awakened in the concentration of iron-ores by magnetism was attested at the fall meeting of the American Institute of Mining Engineers, in New York, by the presentation of four papers, extracts from which will form a large part of this article of the series to which the *Journal* has devoted space. Two of these papers gave data concerning the practical operation of concentrating plants, the third described a special machine, the fourth being devoted to a discussion of the magnetization of brown hematite iron-ore, which was mentioned in the JOURNAL, volume viii., page 236. Before taking up these papers we desire to draw attention to a very important and too often neglected feature of iron-ore concentration, viz.:

THE LOSS OF IRON IN THE PROCESS.

In any method for the concentration of iron-ores the loss of iron in tailings and refuse is an important factor, augmenting in its importance as the percentage of iron in the tails increases, and that in the crude ore or in heads or concentrates decreases. Thus, if the crude iron-ore containing 40 per cent. of iron produces concentrates yielding 60 per cent. of iron, but loses 20 per cent. of iron in the tailings, 2 tons of the crude 40 per cent. material would be necessary to produce 1 ton of 60 per cent. heads and 1 ton of 20 per cent. tails; but if the crude ore yielded but 30 per cent. of iron, and heads and tails similar to the above were produced, 4 tons of the crude material would be required; and 1 ton of 60 per cent. concentrate and 3 tons of 20 per cent. tailings would be produced.

If, however, but 10 per cent. of iron remained in the tailings, $1\frac{2}{3}$ tons of 40 per cent. ore would produce 1 ton of 60 per cent. concentrate and $\frac{2}{3}$ ton of tailings; and, if 30 per cent. crude ore was used, $2\frac{1}{2}$ tons would produce 1 ton of 60 per cent. concentrate and $1\frac{1}{2}$ tons of 10 per cent. tailings. If the separation was carried to a higher and not an impracticable extent, producing heads with 65 per cent. and tails with but 5 per cent. of iron, then but 1.714 tons of 40 per cent. or 2.4 tons of 30 per cent. ore will be required to produce 1 ton of 65 per cent. concentrate, leaving but 5 per cent. iron in 0.714 and 1.4 tons respectively as refuse.

The above are only given as instances to suggest the possible economies in using ores of comparative richness, and of keeping the loss in tailings at a practical minimum. Mr. Henry Helms, of Port Henry, N. Y., suggested a method for determining the number of tons of crude iron required to produce 1 ton of concentrates when the percentage of iron in the crude ore, concentrates and tailings are known. We have arranged this suggestion as a formula, and in presenting it would remind our readers that in it, as in the above calculations, no allowance is made for loss in dust of fine material being blown or washed away. As each method of crushing, breaking, sizing and concentrating effects this item of loss, and as it varies with the physical structure of each ore, no standard can be offered for it.

Formula for calculating the tons of crude ore required to produce 1 ton of concentrate: Let C represent the percentage of iron in the crude ore to be treated; H, the percentage of iron in the finished product or concentrate; T, the percentage of iron in the tailings; X, the number of tons of crude ore necessary to produce one ton of concentrate: then

$$\frac{H - T}{C - T} = X.$$

That is, divide the difference between the iron contents of the crude ore and that of the tailings into the difference between the iron contents of the concentrated ore and that of the tailings. Example: How many tons of iron-ore, yielding 46 per cent. of iron, will be required to produce 1 ton of 68 per cent. concentrates, the tailings carrying 8 per cent. of iron?

$$C = 46\,;\ H = 68\,;\ T = 8$$

$$\frac{68 - 8}{46 - 8} = \frac{60}{38} = 1.58 \text{ tons.}$$

These figures can be verified by multiplying the quantity of crude iron-ore required by its iron contents, deducting from the product the iron in 1 ton of concentrate, and dividing the remainder by the quantity of crude ore less 1 ton as representing the tailings. Thus, taking our first instance, we find that we had:

Two tons of crude iron-ore, containing 40 per cent. of metal, which should give us	0.80 ton of iron.
From this one ton of 60 per cent. concentrate was produced, equivalent to	0.60 " "
Leaving to be accounted for, ...	0.20 " "

As there is remaining one ton of refuse, this is equivalent to 20 per cent. of iron in the tailings. This example also demonstrates that one-quarter of all the iron passed away in the refuse. If, however, this demonstration is applied to a 30 per cent. ore with heads and tails as above, we have:

Four tons of ore at 30 per cent., ...	1.20 tons.
Deduct one ton concentrate at 60 per cent., ...	0.60 "
Leaving as a remainder, ...	0.60 "

Which, divided by 3 tons of refuse, gives 20 per cent. in the tailings. But in this case one-half of all the iron passed away in the tails; in other words, with a 40 per cent. ore, one-quarter (or 10 per cent.) of the iron was lost, and three-quarters (or 30 per cent.) of the iron was saved, while, with a 30 per cent. ore, one-half (or 15 per cent.) of the iron was saved, and one-half (or 15 per cent.) was lost.

This will emphasize the importance of securing tailings low in iron in treating lean ores, for while any unnecessary loss is to be obviated, it is evident that the leaner the crude ore treated, the more important any loss in the tailings becomes.

In discussing the papers above mentioned we shall refer to the formula further, and as an introduction we may properly copy first from the paper presented by Mr. Clemens Jones, Chemist of the Thomas Iron Company, of Hokendauqua, Pa, entitled:

THE MAGNETIZATION OF IRON ORE.

In it, Mr. Jones says:

. . . . "E. S. Dana classifies the natural magnetic minerals in the following order: magnetite, pyrrohotite, franklinite, almandite and minerals 'containing considerable FeO'. Of these, ferroso-ferric oxide, or magnetite, is the type. The proportionate amount of ferrous oxide essential to magnetize the compound, is not established. Ferrous oxide appears to be the only magnetic oxide of iron." . . . .

. . . . "The presence of ferrous oxide in magnetite, whether as a molecular constituent of the compound, or as an independent molecule, seems to communicate the property of magnetism.

"Magnetic oxide of iron, as an artificial product, is considered to result from the direct action of carbon dioxide. In simple determinative blow-pipe work, the reducing-flame alone, or with the aid of alkaline carbonate, produces magnetic particles, or slag-globules, from nearly all iron-minerals.

"Spathic iron-ores have long been known to exhibit, after roasting, magnetic properties—largely due however, to the presence of metallic particles in the roasted ore. Carbon dioxide in the ore is also here considered an important factor." . . . .

. . . . "In nearly all cases of the carbonates of iron, similarly treated, minute globules of metallic iron are contained in the roasted oxide. Furthermore, the expulsion of the carbon dioxide contained in the ore, with contact of carbon at a certain heat, approximates to the conditions of the blast-furnace and favors the reduction of part of the ferrous oxide to the metallic state. The roasted carbonate is always strongly magnetic. But, from the fusible nature of the ore, it is extremely difficult to prevent it from melting during the process, so that nearly all the associated minerals assume the form of 'clinkers' with the ore. This circumstance, together with the expense of crushing the ore sufficiently fine, makes the magnetic separation of iron oxide from the gangue useless or unprofitable in such cases. At Pribram, zinc-blende is sought, and the oxide of iron is a secondary product.

"Hematite is sometimes magnetic and, according to Dana, even magneti-polar—of course, as an incidental occurrence, other than in crystalized specimens. On account of its rarity, as well as the feebleness of the attraction it presents, magnetic hematite is not available for practical purposes. Martite, the isometric ferric oxide which is supposed to be a pseudomorph after magnetite, is attracted by the magnet. The famous ore from Marquette, in the Lake Superior region, abounds with crystals of martite. The ore from the Juragua mines, in Cuba, is in part magnetic. Fayalite, $FeSiO_4$, is magnetic.

"The anhydrous oxides of iron in general are magnetic. Of all the hydrous oxides of iron, not one is magnetic. Brown hematite or limonite (with which göthite and turgite are usually associated), as well as the sub-varieties of hydrous oxides, are non-magnetic, and have no effect upon the magnetic needle. The composition of these oxides is:

Limonite, ...	$2Fe_2O_3+3H_2O$,	or 14.4 parts water.
Göthite, ...	$Fe_2O_3+H_2O$,	or 10.1 " "
Turgite, ...	$2Fe_2O_3+H_2O$,	or 5.6 " "

The importance of the brown hematites is indicated by the amount of such ores consumed in the manufacture of iron in this country and abroad. At one time brown ore was considered indispensable for the production of foundry-iron. Prior to the development of the Lake Superior hematite-regions, brown hematite was the chief ore in use in America. . . . .

. . . . "At present only the better grades are in use, and of these the supply is limited. An investigation of the quality, product and yield of thirty of these mines for the year 1877, undertaken by the writer, showed that over 100,000 tons of this ore, represented by fully 500 samples and analyses, averaged only 37 per cent. of metallic iron. Three-quarters of this was wash-ore. The bulk of this ore, which will fairly represent the majority of this class of mines in the United States, will carry about 34 per cent. of metallic iron in the wash-ore.

"Examination of the iron oxide or ore proper, shows it to contain from 45 to 60 per cent. metallic iron, which in the wash-ore, is, in many instances, reduced below a marketable standard by the admixture of foreign material, chiefly quartz-gravel, slates and clay." . . . .

. . . . "A perfected method of concentrating would mean a survival of the brown-hematite ore as a source of supply.

"On rapidly drying a small sample of limonite over a powerful Bunsen flame, I observed that the smaller particles were magnetized. Could the ore be magnetized? A trial on a more practical scale convinced me that such was the case. I then experimented with several different ores, and found that all the ore-particles were so strongly affected as to permit their complete separation by means of a magnet.

"Naturally, a series of experiments followed, in every instance giving the same results. Attention was then directed toward finding out the conditions as well as the cause of the phenomenon, which led me to the conclusion that the agency of heat in the presence of carbon or carbon dioxide magnetizes the hydrous oxides of iron.

"Further on I will endeavor to qualify this statement; but before beginning its discussion it may be interesting to allude to the practical method and the results, from a commercial standpoint, of some of the experiments.

"The ore is placed in a convenient receptacle and merely roasted by the usual process, using either solid or gaseous fuel. It is undesirable to use much heat, since at a temperature of cherry-red magnetization is fully imparted. The ore is drawn as fast as it reaches this temperature, and is at once ready for magnetic separation.

"In the following list, for obvious reasons, the ores are designated alphabetically:

Iron in Crude Ore. Per cent.	Iron in Concentrate. Per cent.	Iron in Tailings. Per cent.	Of Ore Recovered as Concentrate. Per cent.
A, 40.98	55.88	......	...
B, 34.32	51.72	......	...
C, 38.04	45.24	......	68
D, 40.63	50.04	......	80
E, 20.30	46.32	5.21	55
F, 40.00	53.71	......	65
G, 49	60	......	80
H, 37.06	50	......	65
I, 41.31	52.46	......	85
J, 35.35	55.04	......	60
K, 33.55	48.63	2.56	70
L, 31.31	48.87	5.20	70
M, 39.84	51.29	5.00	78
N, 42.55	55.36	5.76	86
O, 43.84	55.61	......	70
P, 42.96	54.43	......	85

A, *B*, *C*, *D* and *E* are from Pennsylvania, *A* and *B* being wash-ore; *C*, washed and jigged ore; *D*, a fine refuse-sand; and *E*, a similar material; *F*, an ore separated by a Bradford jig; *G*, ore from a Northampton county, Pa., mine; *H*, the tailings rejected by a new jig; *I*, a Pennsylvania ore; *J*, a lot of six cars of ore "condemned" at the works; *K* and *L* are from the same mine in Pennsylvania and impossible to separate by water-jigging; *M* is a mixture of fine ore from Pennsylvania mines; *O*, an ore from Connecticut; and *P*, the fine ore from a large mine in New York.

"The commercial success of treating ores in this way depends on the three items: cost of wash-ore; cost of roasting; cost of concentration. The first may be considered to range from 75 cents to $1.50 per ton, according to locality. The second item will be fully covered by the following statement, taken from the books of a large company exclusively engaged in roasting carbonate ores under the most capable management:

Actual Cost of Roasting and Handling Ore.

PER TON OF ORE.

Roasting Ore.				Handling and Wharf Expenses.			
	1886.	1887.	1888.		1886.	1887.	1888.
Labor, including car- and top-man..	$0.0083	$0.0673	$0.0811	Labor, includ'g engineer & teamster	$0.0514	$0.0439	$0.0720
Supplies..................	0.0024	0.0032	0.0091	Supplies and coal....	0.0177	0.0132	0.0169
Coal.........................	0.0725	0.0750	0.0748	Team........................	....	0.0040	0.0062
				Repairs, etc.............	0.0046	0.0025	0.0029
	$0.1632	$0.1455	$0.1650		$0.0731	$0.0636	$0.0980
Culm, per ton.........	2.00	2.10	2 25				
Average cost. $0.16. About one ton of culm used to 30 tons of ore.				Average....................$0.07 Average total cost... 0.23 Per ton of ore treated.			

"The third item remains to be determined, namely, the cost of concentration, including royalty on the process. This would be 2 cents per ton for concentrating and 25 cents royalty.* Taking as an example ore 'K' in the foregoing list, the cost of the concentrated ore would be as follows:

Cost of 1½ tons ore, ...	$2.25
" roasting, ...	36
" concentration and royalty, ...	27
	$2.88

"At a yield of 48 per cent. iron, this is 6 cents a unit, or, in round numbers, $6.00 to the ton of iron.

"To determine the extent to which the other constituents of the ore were affected, I tested several of the ores, of which two examples will suffice:

Ore E.	Raw.	Concentrates.	Tailings.
Iron, ...	20.30	46.32	5.21
Phosphorus, ...	0.12	0.30	0.09
Ore L.			
Iron, ...	31.31	48.87	5.20
Silica, ...	......	20.25	83.20
Manganese, ...	......	2.20	1.18

* See editor's remarks at end of quotations.

"The surprisingly small amount of ferrous oxide in the magnetized ore leaves the question, whether magnetization is wholly due to its presence, very uncertain. In the list of ores given, G, carrying 60 per cent. of iron, which would be presumably high in ferrous oxide, contains but 3.07 per cent. of ferrous oxide. In some cases the raw ore contains nearly this amount. If the reduction caused by roasting were the sole cause, it would be reasonable to expect the *exterior* of the particle to be the magnetized portion. But lumps of the roasted ore having been broken, small pieces, carefully selected from the inside of the lumps, showed in some cases even more strongly-marked magnetic properties.

"A number of trials with the anhydrous oxides, both red and specular hematites, conducted in precisely the same manner, gave no evidence whatever of resulting magnetization. Crystallized specimens remained likewise unaffected. While the porous character of the brown hematite would permit permeation by reducing gases, it is difficult to understand how such minute deoxidation could magnetize it so strongly. Moreover, the magnetization appears to be permanent. Samples which I have kept for over a year in contact with air have preserved the quality unimpaired. Cooling the red-hot oxide in water affects it somewhat, but it is still capable of responding through that medium to the attraction of a magnetic field.

"The phenomenon seems to be connected with the expulsion of the combined water, and I think is largely dependent on its physical separation from the ore. During the process some of the oxides glow quite strongly. The raw oxide, when heated to redness in a closed glass tube, gives off a small amount of carbonic acid with its water, but is only feebly magnetic. There is no apparent evidence that ferrous oxide is produced by decomposition of the water. The only statement now possible is, that the hydrated oxides of iron become magnetized at a red heat in contact with carbon or carbon dioxide. . . . .

" . . . . It is, of course, a logical deduction that all hydrous ores of iron become magnetic in the blast-furnace, and that at the proper zone even anhydrous hematite does so. Ore concentrated as above described is peculiarly adapted to use in the blast-furnace. Before the ore-particles reach a red heat in the process of roasting, violent decrepitation takes place, thus breaking or splitting up into small fragments all the lumps of ore. This state is most desirable for rapid reduction in the blast-furnace, and, aided by the easy fusibility of the ore, offers the most favorable conditions for "driving" and regularity of work. . . . ."

In giving the average composition of brown hematites we think that Mr. Jones has undervalued the ore, as now prepared, for the large amounts of this class of ore produced in Tennessee, Alabama, Georgia, Texas, Wisconsin and some New England States will bring the average to nearer forty per cent. The allowance of two cents per ton for concentration also evidently only considers the cost of passing the material over or through the concentrating apparatus, without noting the necessity of first comminuting the ore so as to permit of the magnetized portions being separated from the unmagnetized. He has, on the other hand, allowed as royalty an amount greater than should be paid for the use of an apparatus.

We know of several laboratory experiments in magnetizing brown hematite ores, but at present writing have no knowledge of any successful effort to thus produce concentrates commercially.

We may next take up the paper presented by Mr. F. H. McDowell, engineer of the Lackawanna Iron and Coal Company, on

"ORE DRESSING AT THE TILLY FOSTER MINE."

This mine is located in Putnam County, New York, and in the JOURNAL, volume VIII., page 191, we described the concentrating plant connected with it, and to which has since been added Conkling magnetic separators. The work detailed by Mr. McDowell was performed by the mill as we described it and the Conkling separators. These consist of inclined endless belts moved by drums at either end. The ore is fed on to the belt and held to it by magnets placed between the ascending and descending sides. As the material is carried up the incline a stream of water washes the tailings in an opposite direction.

We quote from Mr. McDowell's paper:

. . . . "The local difficulties in the way of this experiment at the Tilly Foster mine were many and great. Only about two-thirds of the waste-dump, resulting from the assortment of previous years, is ore; and this varies in contents of iron from 20 to 28 per cent.

4

The remainder is rock, some (from horses in the vein) showing particles of adhering mineral, some (from the foot-and hanging-walls) entirely barren. The expense of overhauling the dump and removing this rock is by no means the least of the numerous items which must be charged against the concentration. Associated with the lean ore are hornblende and other minerals, carrying iron in the non-magnetic state, which must necessarily be washed away with the tailings. The waste-dump was originally located without reference to future treatment, and can only be handled by means of a locomotive, operating over heavy grades. The mineral is so widely and finely disseminated through the ore that it has been necessary to resort to fine-crushing. This (water being used in the separation) increases the proportion of slimes, which carry off mechanically small particles of mineral. Finally, the lean character of the ore calls for the handling and conveying from the mill of a large bulk of tailings. The process includes crushing the ore by a Blake rock-breaker on the dump; removing it in train-loads to the bins in the mill; passing it under two Ball stamps, provided with screens of $\frac{5}{16}$-inch mesh; elevating it to the Conkling electrical separating-belts; and delivering the concentrates to the cars and the tailings to the settling-reservoirs. . . . . In the face of these changes and the frequent accidents and break-downs incidental to the introduction of new machinery, the mill has been kept running almost continuously ten hours out of the twenty-four, and this has contributed not a little to the success of the undertaking.

"By reference to the accompanying table it will be noted that, during the six months ending July 31, 1890 (the work of January being omitted from the calculation), there were treated 18,058 tons of lean ore, an average of 3009 tons per month. The concentrates produced amounted to 6236 tons, an average of 1039 tons per month. There were required, therefore, 2.89 tons of crude ore to produce one ton of concentrates. The average cost of treatment, including preparing the ore and getting it to mill, and disposing of the tailings (which items make up one-half of the charge for labor), was $2.25 per ton of concentrates. The actual expenses in the mill have averaged $1.63 per ton of concentrates. These figures will be materially reduced when the supply of waste ore from the dump has been exhausted, and a higher grade of lean ore is sent directly from the mine to the mill.

Since the following tables were compiled the figures for August, 1890, have been received. They show 1391 tons concentrates from 3508 tons of crude ore, or 1 : 2.52. The cost per ton of concentrates was $1.89, as compared with $2.39 in July. The reduction of cost is due to the increased output of the mill in concentrates, a larger amount of crude ore having been treated, and also a larger yield obtained per ton of crude ore.

TABLE OF RESULTS FOR SEVEN MONTHS OF 1890.

	January.	February.	March.	April.	May.	June.	July.	Average For 6 mos. Beginning Feb'y 1.
	Tons.	Tons.	Tons.	Tons.	Tons.	Tons.	Tons.	Tons.
Ore used	1719	3007	2278	3104	3619	3120	2930	3009
Concentrates made	445	893	785	1254	1176	1062	1066	1039
One ton of concentrate from	3.75	3.36	2.90	2.47	3.07	2.93	2.61	2.89
Labor, crude to crusher " crushing, and loading cars " repairing cars " " carts " " wheelbarrows " locomotive crew	....	$0.66	$0.59	$0.47	$0.60	$0.63	$0.66	....
" mill work and handling concentrates	....	0.48	0.54	0.37	0.42	0.45	0.44	....
" in ore-bins	....	0.07	0.08	0.06	0.07	0.06	0.06	....
" handling tailings	....	....	0.03	0.01	0.03	0.03	0.02	....
" repairs of mill & mach'ry	....	....	0.17	0.06	0.08	0.10	0.11	....
Analyses	....	....	....	....	....	....	0.03	....
Material for repairs	....	0.40	0.17	0.18	0.19	0.23	0 27	....
Supplies, oil, waste, etc	....	0.16	0.23	0.07	0.21	0.21	0.16	....
Coal used	....	0.66	0.74	0.53	0.55	0.63	0.64	....
Total cost of one ton concentrates	$3.03	$2.43	$2.55	$1.75	$2.15	$2.34	$2.39	$2.25
Cost of labor in one ton "	....	$1.41	$1.40	$0.97	$1.20	$1.27	$1.32	$1.26
Cost to crush and put crude ore in mill	....	$.19\frac{7}{10}$	$.20\frac{3}{10}$	$.18\frac{8}{10}$	$.19\frac{6}{10}$	.21	$.25\frac{3}{10}$	....
Per cent. of iron in crude ore sent to mill	25.13	24.96	28.57	28.21	26.80	28.28	27.55	27.39
Per cent. of iron in concentrates	47.40	50.20	49.54	48.75	50.19	51.04	49.14	....
" " " tailings	....	....	....	....	11.53	10.32	10.21	....
Mill run ... Days	....	$20\frac{8}{10}$	$15\frac{7}{10}$	$23\frac{8}{10}$	$25\frac{3}{10}$	$22\frac{2}{10}$	$22\frac{2}{10}$	....

"The following practical conclusions are suggested by these results :

"1. Unless the location and other conditions are exceptionably favorable, it will not pay to erect works to treat the material of waste-dumps carrying less than 25 per cent. of iron.

"2. Where the lean ore is mixed in connection with shipping-ore, there must be a corresponding increase in the percentage of iron to offset the mining and royalty charges.

"3. Where no shipping-ore is produced, there must be a still further increase in the percentage of iron, to warrant the erection of hoisting-, pumping- and dressing-works. . : . . "

In the record for May, June and July the percentages of iron in crude or heads and tailings are given, and applying the formula above noted, we find a decided difference between the results thus obtained and the actual weights of crude ore fed to the concentrators.

These differences seem to place the formula before referred to in question, but an examination does not confirm this opinion; it rather indicates that there is a serious loss in slimes, or in material carried away by the water, which is not found in the tailings.

Taking the month of May as an instance, we find that one ton of concentrates carrying 50.19 per cent. of iron were obtained from 3.07 tons of crude ore yielding 26.8 per cent. of iron; the tailings carried 11.53 per cent. of iron.

Now 3.07 tons of crude ore 26.8 per cent.	=82.276 units of iron.
Deducting 1 ton of concentrate, . .	50.19 " "
Leaving for tailings, ...	32.086 " "

As there should have been 2.07 tons of tailings, we have

$$\frac{32.086}{2.07} = 15.5 \text{ per cent.}$$

But the table gives only 11.53 per cent. in the tails. Applying the formula to the above results we have

$$\frac{50.19 - 15.5}{26.8 - 15.5} = \frac{34.69}{11.3} = 3.07 \text{ tons.}$$

a result agreeing with the facts.

But if we take the percentages given in the table, we have

1 ton of concentrate, ...	50.19 units of iron.
2.07 tons of tailings at 11.53 per cent., . .	23.87 " "
A total of, ...	74.06 " "

But the 3.07 tons of crude ore at 26.8 per cent. should give 82.28 units, and there is a deficiency of 8.22 units.

These figures demonstrate the correctness of the formula, and locate a loss often unaccounted for in actual practice.

Attention is directed to the fact that at the Tilly Foster plant fine crushing is not resorted to, and as a consequence, a rich concentrate is not obtained from the dense ore. The same remark may be made concerning the third paper, from which we present extracts, viz.:

MAGNETIC CONCENTRATION AT THE MICHIGAMME MINE, LAKE SUPERIOR.

Presented by Mr. John C. Fowle, Manager of the mine, who describes: . . . . "A small crude mill with several series of screens, so as to size the material and treat on the 15 inch drum Wenström separator, the different sizes by themselves. From this mill 11,000 tons of concentrates were shipped in 1889, a part of which, produced from crude 50 per cent. ore, contained 65 per cent. of iron, while the remainder, produced from crude ore running 52 to 54 per cent., carried 60 per cent. iron.

"In view of this satisfactory result in the separation of large and small pieces of ore, my attention was turned to the old dumps about the mine containing a large amount of iron. Finally, crushers were put in to break up this mixed ore and rock; but soft chlorite and hornblende, which were present in large quantity packed the crushers so that they were continually breaking down, and it proved impossible with crushers alone to reduce any considerable quantity of this mixed ore to pass a ⅝-inch screen—the size which had been found in practice most advantageous for securing the maximum of product consistent with sufficient coarseness to suit the condition of the blast-furnace use. We bring the ore screened to this size, up to 61 per cent. in iron, and can raise the iron a few points by carrying less current on the dynamo, which makes the tailings richer.

"The experience at this mill shows that the fine powdered ore and the ore going through a ¼-inch screen, should never be allowed to fall directly on the separator, but should be carried near the separator by a belt on an inclined plane and attracted to the drum of the separator by the electro-magnetic force. It is almost impossible to feed this fine ore directly on to a separator in a sheet sufficiently thin to permit a satisfactory separation, because ore and rock overlying one another are bound together in the drum; but by feeding the fine ore by a belt up to the separator, the mass of material is agitated, and the ore flies to the drum and the rock falls, or remains on the belt. The greater the electric current which is carried on the separator, and the farther away from the separator the crude ore is when it enters the magnetic field, the higher will be the percentage of iron in the concentrates and the lower in the tailings. For instance, crude ore, containing 52 per cent. of iron and 0.224 of phosphorus, when treated directly on the separator, gave concentrates in one case containing 58 per cent. of iron and 0.215 of phosphorus; and in another case 60 per cent. of iron and 0.180 of phosphorus; while the same ore, treated by a belt-feed, not in contact with the separator, gave concentrates containing 67.07 per cent. of iron and 0.060 of phosphorus.

"All of the underground skips that carry lean rock, or 'waste,' are dumped at the shaft-houses upon long, flat screens made of round bar-iron, with bars 1¼ inches apart. The material passing these screens ranges in size from powder up to pieces 1¼ inches, by 3 inches wide, some of which weigh as much as 4 pounds each. These screenings and the lean lump-ore are conveyed to the concentrating-mill over gravity roads. The tracks extend over the top of the mill, and facilitate a convenient delivery into separate large ore-pockets.

On the sides of the mill are skip-roads, by which all the mixed ore from old dumps is hoisted up into the pockets over the crushers. These skip-roads run under the railroad track at a vertical interval which permits the interposition of a pocket holding a railroad carload of ore, 8 tons. By this arrangement the refuse from all parts of the property, when loaded into cars, can be switched at small cost to these pockets and dumped into skips.

"The mill built here is 50 x 60 feet, and has cost $16,000. The machinery includes one 60-horse power engine, two dynamos run by one engine, one boiler, two drums for hoisting skips, one 18 x 24-inch crusher, one 9 x 15-inch crusher, one set of 36-inch rolls, one

36-inch Buchanan separator, one 15-inch Wenström separator, one shaking-screen and a swing-screen.

"The method of treatment in the mill is shown in the accompanying scheme:

Scheme for Concentration.

LUMP-ORE FROM OLD DUMPS AND MINE-WASTE.
Pockets.
18 x 24-inch Crusher.
¾-inch Screen.
Over Screen. | Through Screen.
9 x 15-inch Crusher.
→ Elevated to a ←
→ ⅝-inch Screen.
Over Screen. | Through Screen.
36-inch Rolls. | Swing Screen.
Pockets for different classes, fed separately to —

MINE-SCREENINGS from powder up to pieces 1¼ x 3 inches in size.

WET FINES.
Drier.
Pockets.

^-36-inch Buchanan Magnetic ← Separator
Concentrates. | Tailings.
Shipped. | 15-inch Wenström Magnetic Separator.
Concentrates. | Tailings.
Shipped. | *Discarded*, except in case of coarse mine-screenings, 1¼ x 3 inches in size, which are crushed sized and re-separated.

"The 18 x 24-inch crusher breaks even the largest of our lump-ore; its use saves us much expense in sledging. A coarse screen placed beneath this crusher sifts out the first fines, and delivers all the coarser product to the 9 x 15-inch crusher. Upon passing through this machine, the ore and the first screenings are raised by a link-belt elevator, and delivered to a ⅝-inch screen. Such material as is rejected by this screen, goes to the rolls, and is then again raised by the same elevator and re-sized as before.

"The screen which we employ is a common coal-screen, suspended at top and bottom, and agitated by means of cams. Crushers, rolls, elevator and screen, all work automatically, one man only attending to all.

"The fine ore that passes through the screen is delivered to a swing-screen, from which it drops into several pockets, and is then fed by gravity to the Buchanan separator. This separator also treats, independently of the crushed ore, all the previously-mentioned mine-screenings. These are fed to it direct from storage-pockets, without any preparatory crushing or sizing. The concentrate, or "ore," produced by the separator, is conveyed to railroad cars and loaded for shipment. The tailings are run over our Wenström separator to take out any remaining ore. The Buchanan separator is very well built; it has a much larger capacity than the small Wenström machine, and makes richer concentrates. It carries a current of 23 ampères, and is wound with heavy copper wire. The Wenström machine, on the other hand, can carry only 10 ampères, and is wound with wire of one-third the size.

"A large amount of fine powdered ore and $\frac{1}{4}$-inch screen-ore comes from the Michigamme mine in a wet state, and in order to make a good separation it must be well dried. We have used locomotive sand-dryers, but their capacity is too small. We also use a dryer made by using a half-round cast-iron trough, 16 feet long, with a spiral conveyor running in it to mix up and agitate the wet, fine ore. This trough is set in brick, with a fire-box underneath to burn wood. Wet ore fed into the trough at one end comes out dry at the other end, where it is elevated into the pocket, from which it runs, by gravity, to the separator. This dryer is cheap and easily built, and by using heavy conveyor-flights can be made durable.

"To avoid danger from fire and the cost of a fireman to tend the dryer, we are putting in a revolving conveyor to dry by steam.

"Fine wet ore is to pass by gravity from the feed-hopper through four tubes, which are fixed in an inclined, revolving drum. Live or exhaust-steam enters the drum through a stuffing-box at one end, surrounds the ore-tubes, and thoroughly dries the ore. A slightly different drying-drum has been designed, in which one large tube takes the place of the four small ones. It is expected that this second form will work satisfactorily when the fines are not extremely wet. The drier the material that is to be separated, whether coarse or fine, the greater the capacity of the separator. Up to present writing we have shipped in this season from this mill 8000 tons of concentrates, and are now daily concentrating 180 to 200 tons at a cost, including crushing, hoisting material into mill-pockets, separating ore and loading it into cars, of 18 cents per ton. On some days the concentrates at this mill cost us only 10 cents a ton, delivered in the railroad cars, including all the above-mentioned items of cost.

"The quality of the concentrates is regulated entirely by the electric current and the sizing of the material. Most of our concentrates will pass a $\frac{3}{4}$-inch screen, though some, which are produced from the mine-screenings, are of larger size. These mine-screenings go directly to the separator as 52- or 56-per cent. ore, and leave it as 64- to 65-per cent. ore. Considering their wide range of size, we regard this as very fair work. The waste-rock, resulting from the separation of the screenings, amounts to 10 to 15 per cent. of the total quantity of crude ore. Whenever this waste-product is considered sufficiently rich to warrant further treatment, it is crushed and sized with the lump-ore, and passes through a second separation.

"The proportion of waste which we make from crushed ore is much larger than from the mine-screenings. This only indicates that the lean lump-ore is of poorer grade than the more friable material which breaks into small pieces in mining and forms the screenings. The amount of tailings varies, of course, with the kind of material we are crushing. It fluctuates between 30 and 50 per cent. for most of our separated ore, which runs in size from $\frac{1}{4}$- to $\frac{3}{4}$-inch. The tailings carry from 21 to 24 per cent. of iron. As we break down by far the largest proportion of our rock so as only to pass a $\frac{5}{8}$- to $\frac{3}{4}$-inch screen, it is evident that a lower percentage of iron could easily be obtained by finer crushing. But would such practice be wise policy? I prefer to make a concentrate which sells readily at a good profit, and, to this end, the concentrate must be of such size that it can be worked at low cost and in large quantities by the blast-furnaces.

"Analyses of fine powdered ore up to $\frac{1}{4}$-inch screen-size:

"1. Crude ore.

	Per cent.
Iron,	61.30
Phosphorus,	.131

Concentrates.

	Per cent.
Iron,	64.68
Phosphorus,	.086
Iron,	67.07
Phosphorus,	.080
Iron,	65.99
Phosphorus,	.087
Iron,	63.70
Phosphorus,	.086

"Analyses of concentrates of mixed ores from ¼-inch screens up to pieces weighing 2 to 3 pounds, concentrated at the Michigamme mine, from July 1 to December 1, 1889:

"2. Crude ore, 52 to 54 per cent. iron.
Tailings, 21 to 24 " " "

Concentrates.

	Per cent.
Iron,	62.14
"	63.11
"	61.30
"	62 72
"	64.87
"	63.66
"	63.00
"	63.50
"	63.66
"	60.93
"	62.12
"	60.38
"	60 20
"	60.66

"Analyses of fine powdered ore up to ¼-inch screen-size:

"3. Crude ore, 54 per cent. iron.

Concentrates.

	Per cent.
Iron,	62 72
"	63.50

"4. Crude ore, 56 per cent. iron.

Concentrates.

	Per cent.
Iron,	64.86
"	63.00
"	63.11

"5. Crude ore, 58 per cent. iron.

Concentrates.

	Per cent.
Iron,	64.08
"	63.00
"	63 60
"	66.83

"6. Crude ore, 56.12 per cent. iron.
" " .050 " phosphorus.

Concentrates.

	Per cent.
Iron, ...	62.40
Phosphorus, ...	.040

"The ore was fed directly on to the drum of the separator; by using a belt-feed the concentrates could, in my judgment, have been brought to assay 66 to 67 per cent. iron, and only 0.02 per cent. phosphorus.

"7. Crude ore, 61.30 per cent. iron.
" " .114 " phosphorus.

Concentrates.

	Per cent.
Iron, ...	66.83
Phosphorus, ...	.090"

As a contrast to the two papers just noted, we have one from the pen of Mr. C. M. Ball, of Troy, describing a machine invented by himself and Mr. Sheldon Norton, and giving some experimental results obtained by the machine. We present the following extracts from Mr. Ball's paper on

THE BALL AND NORTON MONARCH MAGNETIC SEPARATOR.

. . . . "The accompanying illustration represents a longitudinal vertical section of the perfected Monarch ore-separator, adapted for separating fine ore.

"The apparatus consists of a partially-closed chest, having an opening at *f*, from the feed-hopper, *h*, through which the ore is delivered to the machine from an ore-pocket or storage-bin provided with means for regulating the flow of ore, so that, when the machine is in operation, the hopper is kept always full. Other openings are provided for the discharge, at *t*, of tailings; at *m*, of middlings; and at *c*, of concentrates; also, at *e*, for allowing free ingress of air to the chest at that point, and at *s*, where a powerful exhaust-fan is connected. The openings at *t* and *m* are kept sealed against ingress of air at those points by means of the hinged and weighted valves, *v*, *v*, which discharge the products from the hoppers, *p* and *k*, continuously, and in the same proportion as received from above, when a sufficient weight has accumulated upon the inside to cause the contents of the hoppers to leak by the valves.

"The machine is also provided, as shown, with two drums, Nos. 1 and 2, turning upon the shafts, *i* and *j*. These shafts, together with the magnets, *a* and *b*, which they also serve to support, stand still, while the drums may be rapidly revolved around the magnets and out of contact therewith.

"It will be noticed that the magnet occupies a sector of the drum, the proportions being such that, approximately, one-third of the periphery of the drum is within the influence of the magnetic field,

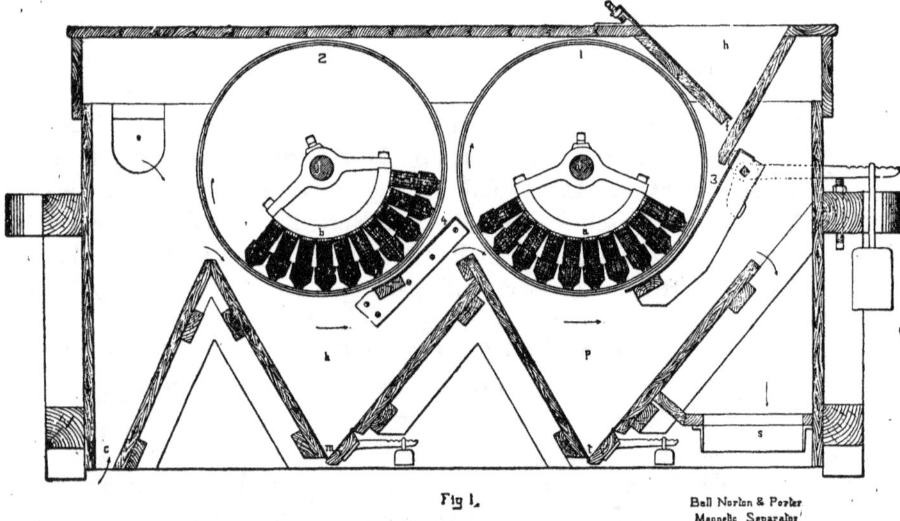

Fig 1.

Ball Norton & Porter Magnetic Separator

while the upper two-thirds is outside of the field and removed from the magnetic influence. The magnet is so constructed as to present a series of poles of alternately opposite polarity near the inner surface of the drum. In accordance with the well-known phenomena of magnetic attraction, which in the case of powerful magnets is exerted at a considerable distance from the magnetic poles, any magnetizable matter brought near the outer surface of the drum, within the arc covered by the magnet, will be powerfully attracted and drawn into firm contact with the outer surface of the drum.

"These drums are composed of a non-metallic and neutral material, such as wood, paper, etc., and they turn in the direction indicated by the arrows near the top of the drums.

"Just below the feed-hopper an apron of neutral metal, No. 3, is arranged, curving downward and forward in the direction of the rotation of the drum, its lower portion describing a short arc concentric to the surface of the drum. This serves as a chute to direct the stream of ore falling from the feed-hopper within the influence of the first two or three poles of the magnet. A similar but somewhat shorter apron, No. 4, is arranged in like relation to the second drum and magnet, *b*.

"When the machine is put in operation the magnets are excited, the drums are revolved in the direction before indicated, the air-current is established through the machine in a direction opposite to that of the rotation of the drums, and ore is supplied through the feed-hopper which is kept always full. The ore passes down the chute under the first drum, and, as soon as it comes within the influence of the magnet, the magnetizable portions are drawn into contact with the drum and through friction upon its surface take on the forward movement of the drum. In accordance with the laws of magnetic induction, a particle of iron brought near a magnet itself becomes a magnet by induction, and the magnetic force tends to bring the longer axis of this induced magnet into a position as nearly as possible parallel to the direction of the magnetic force. So a single particle, or many particles, brought near one pole of a magnet stand on end, as it were; and, in the case of many particles simultaneously influenced by the same pole, they form tufts standing out from the pole, their outer ends repelling each other, but all pointing in the direction of the lines of force towards some focus of opposite magnetic polarity. In conformity with this law, the particles of ore in contact with the drum opposite one of the poles of the magnet stand on end, forming tufts, spreading away from each other at their outer ends. As they are drawn along, however, by friction against the moving drum, when they get to a point midway between two poles they lie down flat against the surface of the drum, and, as they are drawn still further along, they again stand on end—but this time the other end out. So, in passing through the magnetic field, they are tumbled end over end as many times as there are poles in the field. The result is, that every time they are reversed in position opportunity is afforded for any non-magnetic particles of gangue, which may have been entangled with the ore, to fall away from the tufts of magnetite; and this result is still further facilitated by the centrifugal tendency and by the counter-current of air.

"When the ore reaches the limit of the arc covered by the magnetic field it is no longer attracted, and takes on a tangential movement, which carries it away from the drum. It has now, however, passed the edge of the second apron, and, on leaving the first drum, comes within the influence of the magnet of the second drum, where similar operations are repeated, a portion being finally discharged as concentrate at *c*. The function of the second drum and magnet being to differentiate the product from the first drum into two portions, which may be conveniently designated as middlings, discharged at *m*, and concentrate, discharged at *c*. The middlings consist of particles of ore with adhering portions of gangue, which may require a little finer crushing to effect their mechanical liberation; or, they may consist in part of iron compounds having a smaller degree of magnetic susceptibility than the pure magnetite. The separation of the middlings from the mass delivered to the second drum may be effected in two ways: If the drums have the same speed of rotation, a weaker magnetism in the second magnet will allow these less magnetic particles to drop away; or, if the magnets have approximately the same force in the two drums, a higher speed of rotation of the second drum will throw these particles off by reason of the centrifugal force overpowering the centripetal magnetic attraction, the magnet having the smallest influence upon the leaner portions of the mass.

"Modifications of this apparatus have been designed, which adapt it for use in conjunction with a stream of water mingled with the ore, so constituting a wet process.

"In designing this apparatus, it has been assumed that, in order to separate effectively a mass of commingled ore and gangue, containing a large percentage of fine dust, positive and contending forces must necessarily be brought into active play.

"The importance and special efficiency of the means provided for suppressing the movement and discharge, along with the ore, of the finer particles of gangue, and of the means for differentiating the product into several portions graded according to the percentage of iron in each portion, and its magnetic susceptibility, should not be overlooked. It is upon the application of these means that the ability depends to perform the critical operation of converting non-Bessemer into Bessemer ores by the elimination of phosphorus,-sulphur-, and titanium-bearing compounds.

"In the concentration of ores by the dry process the control of the dust developed in the rapid movement of large quantities of ore, is generally difficult. When the ore reaches this machine, the finer portions of the material are as readily and perfectly dealt with and separated as the coarser, the concentrates, however finely pulverized, being delivered from the machine in an absolutely clean condition, and free from non-magnetic dust and the dust carried away by the air-current being also as wholly free from any magnetic material.

"The peculiar action of the ore in passing through the complex or multipolar magnetic field has already been described. Upon the mode of arrangement of the poles the efficiency of the apparatus largely depends. In this machine the magnetism of the field of attraction performs a two-fold function, *viz.*, to attract the magnetizable matter of the ore and, as it is moved through the machine by friction upon the revolving drum, to turn it over and over so as to allow the gangue to fall away, and also to permit the air-current to take effect on all sides of every particle of the ore.

"If the importance of this operation is once recognized, it will be readily understood that there can be only one possible efficient arrangement of the poles, and that it is not a matter of indifference whether they conform to to this or some other plan.

"The positive character of the functions above enumerated renders it possible to make an effective separation under the conditions of a very heavy supply of ore to the machine. The easy working-capacity of a machine having drums of 24 inches diameter and 24 inches working-face is from 15 to 20 tons per hour of ore granulated to pass 16- to 20-mesh screens.

"The power required is from 1 to 1½ horse-power in electricity for each drum, and ½ to ¾ horse-power to drive the machine.

"These machines have been applied to the treatment of a considerable number of ores with results shown in the appended tables.

The most remarkable of these was the conversion of Port Henry Old Bed ore into a Bessemer ore, carrying Fe, 71.10; P, 0.037. This concentration was made from the crude ore, carrying Fe, 58.7, P_2O 2.25; the Bessemer concentrate representing about 65 per cent. of the original mass.

"An earlier type of this machine was in use during nearly the whole of the year 1889, at Benson Mines, St. Lawrence county, N. Y., concentrating the lean magnetite mined there, which in the crude ore carries about 40 per cent. Fe, and is above the Bessemer limit in both phosphorus and sulphur. The concentrated product found a market at Pittsburgh, Pa. The analyses in Table A show the general character of the results; and those in Table B, the results obtained there with the improved machines of the later type.

"The new works at this mine (in course of erection) are provided with crushing-machinery having sufficient capacity to crush daily 800 to 1000 tons of ore, sized to 16-mesh and finer, and the separation of this quantity of ore is to be effected at the outset with three separators of the type represented in Fig. 1, having a working-face on the drums of 24 inches.

TABLE A.

Analyses from Tests with the Ball-Norton Electro-Magnetic Separator, Earlier Form.

MINE.	CRUDE ORE.		CONCENTRATES.		TAILINGS.		
	Fe.	P.	Fe.	P.	Fe.		Per ct. of Original Weight.
Benson Mines, Little River....	45.48	0.158	61.40	0.042	5.60	1.25	29.0
	41.24	0.25	63.31	0.042	5.47	....	....
	40.51	0.19	63.68	0.045	3.87	....	....
	*43.02	0.19	62.10	0.022	....	....	....
Port Henry, New Bed..................	47.70	0.025	70.40	0.011	4.40	0.061	38.0
" " Old Bed, 21.............	58.30	2.18	71.10	0.11	8.70	10.61	19.3
Flagler, Vt., ore	23.74	0.408	64.87	0.084	4.13	0.465	68.44
Chateaugay, N. Y. 16-mesh......	44.56	....	66.69	....	1.26	....	44.0
4-mesh......	33.99	....	59.58	....	10.64	....	52.0
8-mesh......	31.23	....	62.11	....	7.32	....	61.11
16-mesh......	36.23	....	66.98	....	8.12	....	59.0
16-mesh......	†19.60	....	60.48	....	8.96	....	84.375
Tilly Foster..................................	50.25	....	61.12	....	7.88	....	20.0
Beach Glen, N. J., Ore.................	‡50.81	0.042	64.88	0.02	7.06	0.082	25.0

* Average of eight lots shipped from mine. The crude ore contained also: SiO_2, 28.31; S 0,856; the concentrates, SiO_2, 5.94; S, 0.127.

† Tailings from Conkling jigs.

‡ Sulphur as follows: in crude, 0.031; in concentrates, .022; in tailings, 0.02.

TABLE B.

Analyses from Tests with the Ball-Norton Electro Magnetic Separator, Latest Form.

MINE.	CRUDE ORE.		CONCENTRATES.				TAILINGS.		ANALYST.
	Fe.	P.	Fe.	P.	SiO_2	S.	Fe.	Per ct. of Original Weight.	
Benson M'nes Little River, 16-mesh	45.48	....	68.45	0.022	4.78	....	....	....	Donohue.
Benson M'nes Little River, 16-mesh	45.48	....	67.49	0.016	5.64	0.32	....	....	Donohue.
Benson M'nes Little River, 20-mesh	45.48	....	68.62	0.0106	4.18	0.34	....	....	Donohue.
Benson M'nes Little River, 20-mesh	45.48	....	67.81	0.013	5.19	0.27	3.39*	....	Donohue.
Chateaugay tailings......	11.80	....	68.365	0.008	4.31	....	4.33	89.0	Wuth.
Port Henry, New Bed....	47.70	....	70.9	0.0089	1.48	....	....	38.0	Wuth.
" " Old Bed, 21	58.70	2.25	71.1	0.037	....	....	....	20.0	‡
Croton, N. Y., ore..........	†42.99	0.153	69.86	0.021	1.71	0.04	7.95	47.0	Wuth.

Referring to the tabulated result, we regret that in each case the mesh to which the crude material was crushed is not mentioned; it is noticeable that some of the instances in which this is given show that fine comminution increases the percentage of iron in the concentrates—thus we have iron in the Chateaugay ore:

	Crude.	Heads.	Tails.
4 mesh, ...	36.90	59.58	10.64
8 "	31.23	62.11	7.32
16 "	36.23	66.98	8.12
16 "	44.56	66.69	1.26
16 old tailings, ...	19.60	60.48	8.96

As this article has already reached an unusual length we shall not venture any discussion upon Mr. Ball's paper, in the hope that practical commercial operation will before long offer additional data upon which to base conclusions.—EDITOR.

The Population of the United States.

HON. Robert P. Porter, Superintendent of the Census, has issued a bulletin giving the population of the United States, on June 1, 1890, as sixty-two million, four hundred and eighty thousand, five hundred and forty (62,480,540). This is exclusive of white

* The sulphides and hornblende account for more than 2 out of the 3.39 iron in these tailings.

† Sulphur in crude, 0.30.

‡ Witherbee, Sherman & Co.'s chemist.

persons in Indian Territory, Indians on reservations and Alaska. In 1880, the population was fifty million, one hundred and fifty-five thousand, seven hundred and eighty-three (50,155,783), the absolute increase in ten years being twelve million, three hundred and twenty-four thousand, seven hundred and fifty-seven (12,324,757) or 24.57 per cent. In 1880 there were forty-seven divisions covering the entire country, with the exceptions named, and in 1890 there are forty-nine divisions, the additions arising from the opening of Oklahoma and the division of Dakota.

New York heads the list, followed by Pennsylvania, these States occupying the same relative positions as in 1880. In round numbers, New York has a population of six million and Pennsylvania of five and one-quarter million, while in 1880, these two States had five million and four and one-quarter million respectively.

Ohio and Illinois each had between three and four million in 1880, and also in 1890, but they have changed relative positions, Illinois now taking precedence. These two States combined have now seven and one-half million inhabitants.

Five States show between two and three million in 1890, only one of which, Missouri, exceeded two millions in 1880. These five States have an aggregate population of eleven and one-half million, which practically is the same as that of New York and Pennsylvania combined.

Speaking again in round numbers, we find that of the nine States having over two million population in 1890, New York and Pennsylvania together represent 18 per cent. of the population.

Illinois and Ohio together, represent 12 per cent. of the total population.

Missouri, Massachusetts, Texas, Indiana and Michigan together represent over 18 per cent. of the total population.

Making a total for the nine States of over 48 per cent. of the total population or nearly one-half.

Of the remaining States and Territories eighteen have a population of between one and two million, as against fourteen in 1880, three have a population of between one-half and one million in 1890, as against ten in 1880, fourteen have between one hundred thousand and five hundred thousand in 1890 against twelve in 1880 and five had less than one hundred thousand in 1890 against six in 1880.

The Superintendent divides the country into five general divisions, namely:

The North Atlantic, embracing the States from and including Maine, to and including Pennsylvania, which show a population of seventeen million, three hundred and sixty-four thousand, four hundred and twenty-nine, an increase since 1880 of nearly twenty per cent.

The South Atlantic Division embraces all the coast States from Pennsylvania south to and including Florida and also West Virginia. These have a population of eight million, eight hundred and thirty-six thousand, seven hundred and fifty-nine an increase over 1880 of over sixteen per cent.

The Northern Central Division embraces the States north of the Ohio River and westward to the Rocky Mountains, including Missouri and Kansas; the total population of these is twenty-two million, three hundred and twenty-two thousand, one hundred and fifty one, an increase over 1880 of twenty-eight and one-half per cent.

The Southern Central Division embraces the States between the Southern Atlantic Division and the Rocky Mountains, and south of the Ohio River; the population being ten million, nine hundred and forty-eight thousand, two hundred and fifty-three, an increase of twenty-two and three-quarters per cent.

The Western Division includes the balance of the States and Territories except Alaska, and has a population of three million, eight thousand, nine hundred and forty-eight, an increase of seventy and one-quarter per cent.

The Equipment of a Great Steel Works.

A BLAST-FURNACE plant which can produce 2000 tons of pig-iron per day is of itself a marvel, and yet it only forms a portion of one of our large steel rail producing plants. The Edgar Thomson Steel Works at Bessemer, Pa., has a capacity of the above amount, but it is seldom that all the furnaces are operating simultaneously. Recently, however, the average daily output of the pig metal from the plant has been 1550 tons, as set forth in a very interesting and beautifully prepared description of the works issued as a souvenir of the late visit of the European engineers to the United States. This monograph also states that the largest output any one furnace reached in a day was 457 gross tons, the greatest product of one furnace in a week being 2462 gross tons, and in a month one of the stacks produced 10,164 gross tons. Most of this metal is tapped into ladles and carried direct to the Bessemer convertors, the output of Sunday only being cast into chills. These ladles, which each hold 10 tons, are conveyed over a standard-gauge railroad to two "mixers," each having a capacity of 100 tons, and in these the tappings from the various furnaces are thoroughly mixed so as to produce a uniform quality of metal.

The following tabulated statement gives the dimensions of the furnaces forming the plant, the number and size of the fire-brick stoves connected with them, and a statement is also made of the equipment for producing steam and generating blast.

Stack.	Heighth.	Bosh.	Stoves.
Furnace A,	65′	15′	4 stoves 65′ x 15′
Furnace B,	80′	20′	4 " 75′ x 20′
Furnace C,	80′	20′	{ 2 " 75′ x 20′ 2 " 75′ x 21′
{ Furnace D,	80′	23′	6 " 78′-6″ x 21′ }
Furnace E,	80′	23′	1 " 78′ x 20′ }
{ Furnace F,	80′	22′	{ 7 " 78′-6″ x 21′
Furnace G,	80′	22′	
{ Furnace H,	90′	22′	{ 7 " 78′-6″ x 21′
Furnace I,	90′	22′	

Total, 9 Furnaces; 33 Stoves.

To supply blast to these furnaces 26 vertical blowing engines, all of which have blowing cylinders 7 feet in diameter, are in use. The strokes of eleven of these are 5 feet and of fifteen 4 feet. The diameters of the steam cylinders are as follows: Two of 32 inches, thirteen of 35 inches, and eleven of 40 inches.

Steam for these furnaces is generated in 116 boilers, and the equipment is being increased by the construction of 32 additional ones. Of these boilers, 64 now in use and 32 which are building are 54 inches in diameter, 28 feet long with two 18-inch flues. The balance of the boilers being double tier in construction, 28 have upper boilers 50 inches in diameter and 54½ feet long, lower boilers 40 inches in diameter and 44 feet long. Sixteen are nearly the same length; but in twelve of these the diameter of the upper boilers is 42 inches, and of the lower boilers, 32 inches; and four have upper boilers, 36 inches in diameter and lower boilers, 28 inches in diameter. The remaining eight boilers of the equipment have upper boilers, 48 inches in diameter by 32 feet long; lower boilers, 36 inches in diameter by 21½ feet long.

In the steel department are four 10-ton convertors to which the molten pig-iron is taken from the mixers, thus obviating the use of cupolas, although there are five in the building for emergencies. The convertors are blown by three vertical engines with 42″ steam cylinders, 54″ air cylinders and 48″ stroke; one vertical double engine with 36″ steam cylinders, 54″ air cylinders and 48″ stroke. Four No. 7 blowers are in place for the cupolas. Instead of being stripped in the pits, the ingots are pushed out of the molds by means of two hydraulic ingot pushers. The ingots are reheated in nine furnaces using natural gas as a fuel.

The blooming mill consists of a three-high 36″ blooming train and table, which are driven respectively by a 36″ x 72″ and a double 9″ x 12″ engine, a shear, shear-engine 14″ x 24″; also a 3-ton hammer. From the hammer the blooms are carried by a series of driven rollers to a switch, which distributes them to a car running to the five reheating furnaces. From there the blooms are carried to a three-high 24″ rail train, then to a three-high second roughing train, and, finally, to a two-high finishing train of 24″ rolls. The first and second train are run by two 46″ x 60″ and the third by a 30″ x 48″ engine. The mill is also equipped with hot saws operated by a 16″ x 21″ engine and two hot beds.

The finishing department contains the two cold beds, eight straightening and eight drill presses. The daily output of finished rails has averaged 1075 tons, sufficient to lay ten miles of single track, the best record being 1417 tons per day, 7222 tons per week, 30,005 tons per month.

Steam for the steel department is furnished by ninety boilers, seventy of which are finished and twenty building.

32 are	28′ long,	54″ diameter,	with two flues.				
4	"	28′	"	48″	"	"	" "
8	"	28′	"	48″	"	"	four flues.
44	"	28′	"	54″	"	"	two 16″ flues.

Two Heine boilers, 16′ long, two flues, 30″ diameter, with one hundred and thirteen tubes, 3½″ in diameter.

Twenty-five million gallons of water are used per day, 19 pumps supplying the steel department and 24 the blast-furnace plants. Light is furnished by three Brush 65-light dynamos driven by three 11″ x 22″ Buckeye engines running 175 arc lights.

Book Notices.

AMONG those who participated in the late meeting and excursions of the British Iron and Steel Institute was Mr. H. Bauerman, F.G.S., the editor of *A Treatise on Metallurgy of Iron*, which occupies a place in many of the libraries of those interested in the manufacture of iron in the United States. The first issue of this publication, in 1868, has been followed by others, and the demand for the work has been such as to have encouraged the publication, last September, of the sixth edition.

While the book is not written to undertake a detailed technical discussion of the methods and appliances in iron metallurgy, it is of value to any one desirous of obtaining a general knowledge of the physical properties of the materials and the processes used in the manufacture.

WE are in receipt of a copy of the first issue of the *Bibliotheca Polytechnica*, a directory of technical literature, edited by Mr. Fritz von Szczepanski, of St. Petersburg, Russia. It is a classified catalogue of books, annuals and journals published in America, England, France and Germany. It appears to have been compiled with industry, and is a valuable book of reference for publications of a technical nature. The catchwords are given in three languages, English, French and German, so that readers of either nationality can turn to the branch he seeks in the literature of the latest investigations. An enumeration of the technical journals in the three great languages of the world is also given, and the publication of this record of current literature permits of engineers or those interested in the study of technical subjects keeping posted on what is issued.

Copies can be obtained from the International News Company, New York.

www.ingramcontent.com/pod-product-compliance
Lightning Source LLC
LaVergne TN
LVHW010621110826
845149LV00003B/998

9781418186920